How AI Is Transforming the Organization

The Digital Future of Management Series from MIT Sloan Management Review

Edited by Paul Michelman

How to Go Digital: Practical Wisdom to Help Drive Your Organization's Digital Transformation

What the Digital Future Holds: 20 Groundbreaking Essays on How Technology Is Reshaping the Practice of Management

When Innovation Moves at Digital Speed: Strategies and Tactics to Provoke, Sustain, and Defend Innovation in Today's Unsettled Markets

Who Wins in a Digital World? Strategies to Make Your Organization Fit for the Future

Why Humans Matter More Than Ever

How AI Is Transforming the Organization

How AI Is Transforming the Organization

MIT Sloan Management Review

The MIT Press
Cambridge, Massachusetts
London, England

This book was set in Stone Serif and Stone Sans by Jen Jackowitz. Printed and bound in the United States of America.

Library of Congress Cataloging-in-Publication Data is available.

ISBN: 978-0-262-53839-8

10 9 8 7 6 5 4 3 2 1

Contents

Series Foreword

Books in the Digital Future of Management series draw from the print and web pages of *MIT Sloan Management Review* to deliver expert insights and sharply tuned advice on navigating the unprecedented challenges of the digital world. These books are essential reading for executives from the world's leading source of ideas on how technology is transforming the practice of management.

Paul Michelman
Editor in Chief
MIT Sloan Management Review

Introduction: How AI Is Transforming the Organization

Sam Ransbotham

We have repeatedly seen decades-long cycles of promise and disillusionment around artificial intelligence. While one might thus wonder if the current interest will fade, it's unlikely. Today, AI is a general-purpose technology that businesses can use to create value. Many organizations are already doing so.

MIT Sloan Management Review and Boston Consulting Group's collaborative research on AI found that organizations creating value from AI are large and small, traditional and digital natives, and scattered across the globe and across industries.

Consider Airbus. The company wanted to build its new A350 aircraft more quickly than it had built previous models without compromising quality. Airbus knew that workers learned as they gained experience, a crucial step in fabricating a brand-new aircraft. But spreading individual workers' knowledge and experience to the other people involved is an inherently slow process.

To disseminate organizational learning more quickly, Airbus turned to AI. The company used AI along with its existing data and analytics infrastructure to capture raw information, to structure that information, and to help workers access it. Importantly,

Airbus didn't start out with a goal of using AI—the company used AI because it was the right tool to enable its employees to learn faster.[1]

Consider Fidelity. Financial services companies must balance helping their customers manage their assets with ensuring those assets remain secure from nefarious actors. Unfortunately, this results in customer service interactions that usually begin by challenging customers to prove their identity. By design, unknown callers are considered fraudulent until proven otherwise. Given how routine and irritating this interaction is, authentication seems like a good candidate for AI.

A knee-jerk example of using AI would be to replace customer service agents with systems designed to mimic human customer service. (Think Agent Smith from *The Matrix*—similarly cheap to replicate, but hopefully with a friendlier disposition.) Instead, AI at Fidelity works in the background. It analyzes voice identification, call latency, and vocal stress to assist human customer service agents in authenticating customers. Fidelity is replacing some of the customer service agents' tasks—not the agents themselves. The tasks they are replacing? The ones the humans don't want to do (such as authenticating callers), can't do well (such as detecting millisecond-long call latency discrepancies), or could use a bit of help with (such as recognizing speech stress patterns). By using AI in the background, Fidelity can start customer service on a positive note rather than an adversarial one while simultaneously reducing call authentication time.[2]

Consider Airbnb. The demand for private property rentals has grown rapidly, with Airbnb leading the online marketplace. However, potential renters face considerable search costs when narrowing down their options. Some attributes of property listings are easy to search for (such as the number of bedrooms and square footage), whereas others are unstructured (such as pictures and text description).

Airbnb turned to AI to extract such unstructured attributes from property photos and other sources to help potential renters identify other, similar properties. As a result, Airbnb is better able to serve its customers' needs and thus improve customer satisfaction with Airbnb's services.[3]

If you find these examples intriguing and you accept the premise that AI can add value to your organization, what should you do next? For starters, read this book.

Part I focuses on the crucial role of people in this technology transformation: How do people and AI interrelate? Part II addresses some of the difficult strategic and operational choices at the top levels: How should executives manage with AI? Part III examines the increasingly important trust and ethical issues that come with this progress in technology: Now that we *can* use AI, how *should* we be using it? Last, Part IV examines how a robot's behavior can positively influence its human colleagues by working *with* us.

Several themes are evident in these chapters. Or, perhaps more accurately, several *un*-themes—the themes are defined more by what they are *not* than what they *are*.

AI in Business Isn't Confined to the Digerati – It's for All Organizations

It is no surprise to find leading technology companies applying AI in their businesses. These digital natives have the data that AI algorithms require as well as the talent to create value with AI. This book doesn't ignore the prominent roles of FAANG (Facebook, Apple, Amazon, Netflix, and Google) companies and their fellow digital denizens, but it also discusses many examples outside of the FAANG cohort. Companies across the board are training workers in the new skills they need to be competitive, including SAP and Walmart in Jacques Bughin's "Preparing for the Coming Skill Shifts" and KFC and Fidelity in Lynda Gratton's "The Challenge of Scaling Soft Skills." Honeywell, Cannondale, and DHL are incorporating data from new AI-based sources, as Joe Biron and Jonathan Lang show in "Unlocking the Value of Augmented Reality Data." Thomas H. Davenport and Vivek Katyal's "Every Leader's Guide to the Ethics of AI" notes that Farmers Insurance established not one but two AI-related governance boards. While the digerati are important, AI is by no means their sole provenance.

AI in Business Isn't Tomorrow – It Is Now

More than 90% of organizations expect new business value from AI implementations in the coming five years, according to *MIT Sloan Management Review* and BCG's research. Organizations that are lagging already feel some urgency and competitive pressure.

R. Edward Freeman and James R. Freeland argue in "The Time for Retraining Is Now" that companies and governments need to engage in AI efforts so that we, not the robots, will be in charge of our future. Perhaps surprisingly, rather than being reluctant to adjust, the people most affected by AI-related change may be readier to get going than their companies are, as Eva Sage-Gavin, Madhu Vazirani, and Francis Hintermann find in "Getting Your

Employees Ready for Work in the Age of AI." Without motivated and knowledgeable leaders, Davenport and Janet Foutty warn in "AI-Driven Leadership," companies that fail to employ AI may miss out on evolutionary progress today—and, more importantly, on future revolutionary changes. And in "Using AI to Enhance Business Operations," Monideepa Tarafdar, Cynthia M. Beath, and Jeanne W. Ross explore why deep business domain proficiency is vital to gaining competitive advantage.

AI in Business Isn't about Humanoids – It's about Humans

News stories seem to gravitate toward AI advances that mimic humans. Robots and Turing tests capture headlines and page views, but there are few examples in this collection of organizations getting value from any AI that looks like or mimics humans. Mimicking human beings is extraordinarily difficult; and, frankly, organizations already have access to resources with human capability—actual humans.

Instead, AI can complement skills that people have. Developing these complementarities will require overcoming numerous difficulties, as Erik Brynjolfsson, Daniel Rock, and Chad Syverson discuss in "Unpacking the AI–Productivity Paradox." I argue in "Don't Let Artificial Intelligence Supercharge Bad Processes" that the best business use of AI involves figuring out better ways to work rather than automating existing human processes. It will involve cooperating with AI, not relinquishing control to it, agree Gary Hoffman in "Building a Robotic Colleague with Personality" and Berkeley J. Dietvorst in "When People Don't Trust Algorithms." Like most technology-focused changes, AI ultimately is about people.

AI in Business Doesn't Have a Clear Prescription – It Requires Adaptation

Some might wish this collection could provide step-by-step guidance to guarantee AI success: Do steps 1, 2, 3, and it is as easy as A, B, C. However, organizational experiences show the process is more nuanced than this. Considerable uncertainty remains about how organizations will directly use AI, and there is even more uncertainty about its indirect effects on other organizations, industry, and society.

As a result, these chapters aren't filled with breathless hype, but instead show a frank appreciation of the risks that AI may initially exacerbate, as in Chris DeBrusk's "The Risk of Machine Learning Bias (and How to Prevent It)." Managing through these risks shows promise, however. Megan Beck and Barry Libert examine how increased transparency could help improve gender inequality in "Could AI Be the Cure for Workplace Gender Inequality?" and Paul R. Daugherty, H. James Wilson, and Rumman Chowdhury explore how AI could help develop a more inclusive world in "Using Artificial Intelligence to Promote Diversity."

Even here, we find nuance. Although AI may help organizations improve in some situations, David Kiron and Gregory Unruh argue in "Even If AI Can Cure Loneliness, Should It?" that its greater societal ramifications may require regulation, while Julian Friedland's "AI Can Help Us Live More Deliberately" warns of the dangers of relying too much on AI at the expense of our own personal agency.

Importantly, managing through all these uncertainties as AI matures is a decidedly human task. Considerable attention is required to nudge simple AI systems to productive use, even within a narrow context. Each chapter speaks to the continued need for human oversight as the scope of current AI-based systems bumps against the limits of their design.

This collection contains concrete examples of how organizations are creating business value with AI today and offers guidance to organizations that have not made as much progress. And it also reveals how AI may transform organizations of the future.

The question isn't whether AI can help, but where and how much help it can give. *MIT Sloan Management Review* and BCG's 2018 Executive Study and Research Report on AI found that only 18% of organizations have an understanding and substantial adoption of artificial intelligence. From 2017 to 2018, however, the vast majority (88%) of those organizations further increased their investments in AI—they aren't waiting for others to catch up.

It is clear that many organizations need to get going. We hope this collection shows the way.

I

People

1

AI-Driven Leadership

Thomas H. Davenport and Janet Foutty

Many companies are experimenting with AI on a small scale, and a few have made a commitment that their organizations will be "AI first" or "AI-driven." But what does this mean? What is AI doing or leading, and, in particular, what is the role of leadership in making organizations AI-driven?

We see a lot of confusion around opportunity and action. In the 2018 Deloitte Global Human Capital Trends survey and report of business and HR leaders, 72% indicated that AI, robots, and automation are important—but only 31% felt their organizations were prepared to address strategy to implement these technologies.

There aren't that many AI-driven companies yet, and while we have seen a few examples outside of tech, most of the ones that do exist are technology companies. That makes it a little speculative to define the traits leaders need to help move their organizations to an AI-first future. Still, it's useful to begin to develop a framework for what we already know about what it takes to be an AI-driven leader.

The Seven Attributes of AI-Driven Leaders

Some of the necessary traits to become AI-driving leaders are simply logical. We also have surveys to draw from of organizations that are active in the AI field. Parsing this material, we have identified seven attributes of leaders at the AI forefront:

They learn the technologies. It helps a lot in leading with AI to know what AI is and does. Yes, this sounds obvious, but by tradition, leaders outside of IT fields have rarely felt the need to understand technologies beyond the surface level. AI is different. It is not just one technology, but many—each with its own application types, stage of development, strengths, and limitations. Some, like robotic process automation, are relatively easy to implement—even for non-IT people—and provide rapid ROI. Others, like deep-learning neural networks, are much more complex and cutting-edge, requiring high levels of expertise. AI-driven organizations typically want to explore a wide variety of technologies, and leaders need to know enough about them to be able to weigh in on which ones will be most critical to their organization's success.

They establish clear business objectives. As with any other technology, it's important to have clear objectives for using AI. Is there a service that could be easier for customers to use? Are there particular business processes in which knowledge isn't effectively distributed? Are some types of decisions not employing data in the way they should? AI can address any of these issues, but probably not all at once, so leaders need to make some choices. The 2017 Deloitte State of Cognitive survey of US executives with a high level of AI awareness and operating within the companies that are most aggressively adopting AI and cognitive technologies found that the most popular objectives involved using AI to improve existing products and services, make better decisions, create new products, and optimize business processes. Of course, there are choices that need to be made deeper within each of those categories. These decisions are appropriate for senior managers to own and make.

They set an appropriate level of ambition. Some organizations have difficulties in pulling off highly ambitious goals, and "moon shots," not surprisingly, don't always reach the moon—and may set back AI initiatives overall if they fail. The alternative is to undertake a series of less ambitious projects—often called the low-hanging fruit. Even at Amazon.com, one of the most technologically sophisticated organizations in the world, CEO Jeff Bezos noted in his 2017 letter to shareholders that while the company is undertaking some highly ambitious projects like intelligent drones, the bulk of its machine learning efforts are devoted to "quietly but meaningfully improving core operations." While less transformational individually, a series of such projects can add up to major change in a product or process. We generally feel that most companies will be better off with a series of less ambitious AI projects, although there may be some circumstances in which a large-scale goal is warranted.

They look beyond pilots and proofs of concept. Our research suggests that AI projects thus far are heavily weighted toward pilots. But to improve productivity and achieve the needed ROI, leaders need to push their companies to scale up these projects to full production status. This means identifying process improvements before applying technology and figuring out how to integrate AI technologies with existing applications and IT architectures. These are not easy tasks. AI-driven leaders must help to assess the potential for full-scale implementation before embarking upon pilot projects.

They prepare people for the journey. Most AI projects will involve augmentation—smart people working in collaboration with smart machines—rather than large-scale automation. That means that employees will have to learn new skills and adopt new roles, which won't happen overnight. Good leaders are already preparing their people for AI by developing training programs, recruiting for new skills when necessary, and integrating continuous learning into their models—and in fact, the 2018 Digital Business report by Deloitte and *MIT Sloan Management Review* shows that businesses are putting an increased focus on continuous learning. Bank of America's Technology and Operations function developed a series of online education programs for its over 90,000 employees that addresses some of the skills needed for work alongside the bank's chatbot, Erica, and other AI applications. Deloitte has focused on making its professionals tech-savvy—and not just those who work in tech consulting—with an understanding that in an AI-oriented business environment, virtually every Deloitte employee will need to understand how technology works and fits with their jobs.

They get the necessary data. AI-driven leaders know that data is their most important asset if they want to do substantial work in AI. Leaders planning to use machine learning to predict what their customers will buy under what circumstances, for example, need high-quality data on what different types of customers have bought in the past. Leaders of health care organizations that want to use deep-learning models to analyze medical images need a lot of images with labeled outcomes from which the system can learn. Many organizations will need to turn to external data to augment their internal sources, while others will need to improve data quality and integration before they can use it with their AI projects. In other words, AI-driven leaders need to start now, if not yesterday, to improve their data.

They orchestrate collaborative organizations. C-level executives—CEOs and heads of operations, IT, HR, marketing, and the like—are not often known for collaborating closely on initiatives involving technology. But these groups need to work together in AI-driven organizations to establish priorities, determine the implications for technology architectures and human skills, and assess the implications for key functions such as marketing and supply chain. If AI is viewed as an important catalytic technology, there is no reason why senior executives should not collaborate in its leadership even if the organization's near-term plans for AI may be relatively modest—because they won't be modest forever. Deloitte has referred to this approach as "symphonic leadership," with players working in concert like an orchestra. Many companies that develop AI projects say they employ agile methods. In AI initiatives that benefit from agile methods, these executives must be involved stakeholders, providing input on goals and deliverables and the impact that the initiatives will have on the business. Together, agile and symphonic leadership teams will not only facilitate progress in AI but also communicate to the organization that a new way of working and managing is being adopted.

Evolutionary Today, Revolutionary Tomorrow

Becoming an AI-driven leader may not be easy for many executives. It's not yet common for executives to even understand the technologies: The 2017 Deloitte survey of executives with high AI awareness found that 37% of the respondents said the primary AI challenge is that "managers don't understand cognitive technologies and how they work."

Leaders will have to work harder to realize the benefits of AI. In most organizations today, AI is a source of evolutionary benefits, but we are confident that over the long run it will be a revolutionary force. AI will lead to changes in how work is done, how decisions are made, and how organizations harness knowledge and information to achieve goals.

There should be little doubt that companies aspiring to be AI-driven need themselves to be led by leaders who are AI-driven—meaning motivated, knowledgeable, and engaged with regard to this powerful tool.

2

Could AI Be the Cure for Workplace Gender Inequality?

Megan Beck and Barry Libert

Many researchers are reporting, and our research confirms, that artificial intelligence will reshape our economy—and the roles of workers and leaders along with it. Jobs that don't disappear will see a significant shift as the tasks that are easily and inexpensively accomplished by robots become automated. The work that remains will very likely focus on relating. To adapt and prosper, the smart worker will invest in human relating skills—empathy, compassion, influence, and engagement. For simplicity, let's call these emotional quotient (EQ) skills. These are skills in which women commonly excel.

Gender differences are a sensitive topic, and we address them in this article with trepidation. There is a fine line between understanding commonalities and stereotyping, and the debate about nature versus nurture is robust. But whether you believe that men and women, on average, have different types of brains (as Simon Baron-Cohen, a British clinical psychologist and professor of developmental psychopathology at the University of Cambridge, has theorized) or that gender differences are a result of cultural norms and conditioning (as numerous other studies have explored), the real-world results are similar: Men and women, on average, excel in different dimensions and take on different roles in the workforce. By no means does that suggest that men and women are not equal—just different.

It is clear that men have quite an advantage in the working world—just check out the latest research by McKinsey & Co. on gender equality in the workplace. Men have greater representation among leadership roles, greater presence in higher-paid industries, hold nearly 80% of board seats, and earn higher compensation on average, even for the same jobs.

We believe that AI has the ability to help level the playing field. It will do so, we think, by replacing many roles and functions where men typically dominate.

Jobs That Currently Demand High EQ Are Dominated by Women

An examination of common occupations by gender in the US by the Department of Labor reveals some unsurprising data. Women predominate in jobs that involve relating, caretaking, and providing services, making up more than 80% of the country's school teachers, nurses and home health aides, social workers, and secretaries and administrative assistants. Men outweigh women in fields that tend to be physical, STEM- and finance-related, and more isolated rather than relational, such as truck drivers, janitors, laborers, and software developers. Men are also better represented in higher-paying, often analytical fields, such as law, medicine, and engineering.

One perspective on the ways that different skill sets play out at work is the empathizing–systemizing theory, which measures people's inclinations to empathize (identify, understand, and respond to the mental states of others) and to systemize (analyze, understand, and predict system). According to Baron-Cohen, the theory's author, women score higher on empathizing and men higher on systemizing. A recent Korn Ferry report aligns with this point of view: It found that women score higher than men on 11 out of 12 key emotional intelligence competencies. These include demonstrating empathy, conflict management, and coaching/mentoring.

EQ Is Likely to Become a Critical Job Differentiator in More Roles

Differences in current skills and roles mean that the evolving AI economy is going to affect men and women differently.

We all know that changes due to AI are imminent, and that some roles will likely disappear over the next decade. This will not be limited to any particular industry or pay grade. Robots will replace not only truck drivers and stock pickers, but also radiologists, consultants, and financial planners (all of which are traditionally male-dominated roles).

The jobs, or the parts of jobs, likely to have more staying power in the AI economy are those that rely more heavily on EQ—abilities such as empathy, persuasion, and inspiration. AI may determine that your radiology scans indicate cancer, but a human will likely sit down with you and help create a treatment plan that suits your goals and lifestyle. AI may suggest operational improvements within a company, but a human will be more effective at persuading the leadership team to tackle the problem. Chinese technologist Kal-Fu Lee predicts that AI will probably wipe out 50% of jobs within a decade, but adds that nothing can replace human-to-human interaction: "Touching one's heart with your heart is something that machines, I believe, will never be good at," he told CNBC.

Research has suggested that these relating skills, where men lag behind women, will put men at a workplace disadvantage in the AI economy. They won't be as successful as women unless they embrace these differentiator skills of empathizing, mentoring, and engagement.

Three Steps to Prepare for the EQ Revolution

Given this prognosis, all of us—men, women, and the organizations we work with—need to pay real attention to these often-neglected EQ skills.

Although we tend to think of relating skills as innate and static, this is incorrect. Just like any job skill, a person's emotional intelligence can be improved with some effort. Here are three steps to get started:

Figure out what you're working with. What is your EQ baseline? Many sharp, effective people have low EQ but have no idea they need to improve because they simply haven't paid attention to the subtle indicators from their peers. Most of us are very hesitant to criticize someone's interpersonal skills directly because such feedback can be perceived as an unwelcome critique. This means that you should pay attention to couched feedback you've been given, especially comments along the lines of, "You are difficult to work with," "You are too argumentative," "You need to do a better job 'reading the room.'"

Admit to yourself the importance of EQ. Emotional intelligence has been undervalued in the marketplace since . . . forever. Although every job has an EQ component, employees and managers are more often trained and assessed on systemizing skills—perhaps because they are simpler to measure. For example, doctors are well trained on identifying and treating disease. But they are not well trained on personalizing treatment to suit a patient's preference and lifestyle, nor on influencing patients to take steps such as changing diet or exercise. If you want to grow your EQ, you must first change your mental model about what is important in your work. Is getting the diagnosis right the most important measure of success? Or is it actually improving someone's health? Recognize that making an impact on the world almost always involves human interaction.

Practice and train your EQ. Research shows that attention and training programs can affect one's emotional intelligence. Identify the parts of your job that allow you to practice understanding, coaching, encouraging, and influencing others—these are the parts of your role likely to persist over the next decade—and direct your energy to these interpersonal opportunities. Find a coach who will give you honest feedback and mentoring, or find a training program. We naturally take these steps with many job skills, but are hesitant to do so with EQ for two reasons: None of us want to admit our EQ needs work, and we have the idea that our EQ is inborn and unchangeable. We are wrong on both accounts.

Whether it is genes or training that inclines women to empathize, relate, and engage more than men is irrelevant. As AI-based tools become integrated into roles across levels and industries, these soft skills will become more important for earning hard dollars.

Companies and organizations need to be aware of this shift in job skills, as it will affect hiring, managing, and training employees. Those who can't adjust will see their skills become irrelevant, from the boardroom to the manufacturing floor. There are many things that people will not be able to do as effectively as the robots that are moving into our workplaces, so it's time to focus on what people can do best—understanding and relating to each other.

3

Using Artificial Intelligence to Promote Diversity

Paul R. Daugherty, H. James Wilson, and Rumman Chowdhury

Artificial intelligence has had some justifiably bad press recently. Some of the worst stories have been about systems that exhibit racial or gender bias in facial recognition applications or in evaluating people for jobs, loans, or other considerations.[1] One program was routinely recommending longer prison sentences for blacks than for whites on the basis of the flawed use of recidivism data.[2]

But what if instead of perpetuating harmful biases, AI helped us overcome them and make fairer decisions? That could eventually result in a more diverse and inclusive world. What if, for instance, intelligent machines could help organizations recognize all worthy job candidates by avoiding the usual hidden prejudices that derail applicants who don't look or sound like those in power or who don't have the "right" institutions listed on their résumés? What if software programs were able to account for the inequities that have limited the access of minorities to mortgages and other loans? In other words, what if our systems were taught to ignore data about race, gender, sexual orientation, and other characteristics that aren't relevant to the decisions at hand?

AI can do all of this—with guidance from the human experts who create, train, and refine its systems. Specifically, the people working with the technology must do a much better job of building inclusion and diversity into AI design by using the right data to train AI systems to be inclusive and by thinking about gender roles and diversity when developing bots and other applications that engage with the public.

Design for Inclusion

Software development remains the province of males—only about one-quarter of computer scientists in the United States are women[3]—and minority racial groups, including blacks and Hispanics, are underrepresented in tech work, too.[4] Groups like Girls Who Code and AI4ALL have been founded to help close those gaps. Girls Who Code has reached almost 90,000 girls from various backgrounds in all 50 states,[5] and AI4ALL specifically targets girls in minority communities. Among other activities, AI4ALL sponsors a summer program with visits to the AI departments of universities such as Stanford and Carnegie Mellon so that participants might develop relationships with researchers who could serve as mentors and role models. And fortunately, the AI field has a number of prominent women—including Fei-Fei Li (Stanford), Vivienne Ming (Singularity University), Rana el Kaliouby (Affectiva), and Cynthia Breazeal (MIT)—who could fill such a need.

These relationships don't just open up development opportunities for the mentees—they're also likely to turn the mentors into diversity and inclusion champions, an experience that may affect how they approach algorithm design. Research by sociologists Frank Dobbin of Harvard University and Alexandra

Many of the data sets used to train AI systems contain historical artifacts of biases, and if those associations aren't identified and removed, they will be perpetuated.

Kalev of Tel Aviv University supports this idea: They've found that working with mentees from minority groups actually moves the needle on bias for the managers and professionals doing the mentoring, in a way that forced training does not.[6]

Other organizations have pursued shorter-term solutions for AI-design teams. LivePerson, a company that develops online messaging, marketing, and analytics products, places its customer service staff (a profession that is 65% female in the United States) alongside its coders (usually male) during the development process to achieve a better balance of perspectives.[7] Microsoft has created a framework for assembling "inclusive" design teams, which can be more effective for considering the needs and sensitivities of myriad types of customers, including those with physical disabilities.[8] The Diverse Voices project at the University of Washington has a similar goal of developing technology on the basis of the input from multiple stakeholders to better represent the needs of nonmainstream populations.

Some AI-powered tools are designed to mitigate biases in hiring. Intelligent text editors like Textio can rewrite job descriptions to appeal to candidates from groups that aren't well-represented. Using Textio, software company Atlassian was able to increase the percentage of women among its new recruits from about 10% to 57%.[9] Companies can also use AI technology to help identify biases in their past hiring decisions. Deep neural networks—clusters of algorithms that emulate the human ability to spot patterns in data—can be especially effective in uncovering evidence of hidden preferences. Using this technique, an AI-based service such as Mya can help companies analyze their hiring records and see if they have favored candidates with, for example, light skin.

Train Systems with Better Data

Building AI systems that battle bias is not only a matter of having more diverse and diversity-minded design teams. It also involves training the programs to behave inclusively. Many of the data sets used to train AI systems contain historical artifacts of biases—for example the word *woman* is more associated with nurse than with doctor—and if those associations aren't identified and removed, they will be perpetuated and reinforced.[10]

While AI programs learn by finding patterns in data, they need guidance from humans to ensure that the software doesn't jump to the wrong conclusions. This provides an important opportunity for promoting diversity and inclusion. Microsoft, for example, has set up the Fairness, Accountability, Transparency, and Ethics in AI team, which is responsible for uncovering any biases that have crept into the data used by the company's AI systems.

Sometimes AI systems need to be refined through more inclusive representation in images. Take, for instance, the fact that commercial facial recognition applications struggle with accuracy when dealing with minorities: The error rate for identifying dark-skinned women is 35%, compared with 0.8% for light-skinned men. The problem stems from relying on freely available data sets (which are rife with photos of white faces) for training the systems. It could be corrected by curating a new training data set with better representation of minorities or by applying heavier weights to the underrepresented data points.[11] Another approach—proposed by Microsoft researcher Adam Kalai and his colleagues—is to use different algorithms to analyze different groups. For example, the algorithm for determining which female candidates would be the best salespeople might be different from the algorithm used for assessing male candidates—sort of a digital affirmative action tactic.[12] In that scenario, playing a team sport in college might be a higher predictor of success for women than for men going after a particular sales role at a particular company.

Give Bots a Variety of Voices

Organizations and their AI system developers must also think about how their applications are engaging with customers. To compete in diverse consumer markets, a company needs products and services that can speak to people in ways they prefer.

In tech circles, there has been considerable discussion over why, for instance, the voices that answer calls in help centers or that are programmed into personal assistants like Amazon.com's Alexa are female. Studies show that both men and women tend to have a preference for a female assistant's voice, which they perceive as warm and nurturing. This preference can change depending on the subject matter: Male voices are generally preferred for information about computers, while female voices are preferred for information about relationships.[13]

But are these female "helpers" perpetuating gender stereotypes? It doesn't help matters that many female bots have subservient, docile voices. That's something that Amazon has begun to address in its recent version of Alexa: The intelligent bot has been reprogrammed to have little patience for harassment, for instance, and now sharply answers sexually explicit questions along the lines of "I'm not going to respond to that" or "I'm not sure what outcome you expected."[14]

Companies might consider offering different versions of their bots to appeal to a diverse customer base. Apple's Siri is now

Studies show that both men and women tend to have a preference for a female digital assistant's voice, which they perceive as warm. But are female "helpers" like Alexa reinforcing gender stereotypes?

available in a male or female voice and can speak with a British, Indian, Irish, or Australian accent. It can also speak in a variety of languages, including French, German, Spanish, Russian, and Japanese. Although Siri typically defaults to a female voice, the default is male for Arabic, French, Dutch, and British English languages.

Just as important as the way they speak, AI bots must also be able to understand all types of voices. But right now, many don't.[15] To train voice recognition algorithms, companies have relied on speech corpora, or databases of audio clips. Marginalized groups in society—low-income, rural, less educated, and non-native speakers—tend to be underrepresented in such data sets. Specialized databases can help correct such deficiencies, but they, too, have their limitations. The Fisher speech corpus, for example, includes speech from non-native speakers of English, but the coverage isn't uniform. Although Spanish and Indian accents are included, there are relatively few British accents. Baidu, the Chinese search-engine company, is taking a different approach by trying to improve the algorithms themselves. It is developing a new "deep speech" algorithm that it says will handle different accents and dialects.

Ultimately, we believe that AI will help create a more diverse and better world if the humans who work with the technology design, train, and modify those systems properly. This shift requires a commitment from the senior executives setting the direction. Business leaders may claim that diversity and inclusivity are core goals, but they then need to follow through in the people they hire and the products their companies develop.

The potential benefits are compelling: access to badly needed talent and the ability to serve a much wider variety of consumers effectively.

4

Preparing for the Coming Skill Shifts

Jacques Bughin

About 80% of US and European CEOs surveyed by McKinsey say they worry about ensuring that their companies have the right skills mix to thrive in the age of AI and automation. Those leaders come from a variety of industries, and they're smart to be thinking about talent at a strategic level.

Given the complexities of implementing the new technologies, companies will, of course, need people who can design the right algorithms and interpret the data. But they'll also need so-called "softer" skills to do the work that machines aren't capable of doing. Our research suggests that demand for social and emotional skills will grow by about one-quarter by 2030, and we also see a clear shift toward higher cognitive skills, including creativity and complex information processing.

An HR tool kit for shaping a workforce already exists: Companies can retrain people, redeploy them to make the best use of skills available, contract work out, hire new people, and release those who do not meet the organization's needs. But the external labor market can do only so much to address the anticipated shifts in demand, the pace of which will accelerate over the next decade, according to our research. If the volume of companies seeking to hire for the necessary technical and soft skills rises too rapidly, full-time salaries and contractor rates will skyrocket, and most organizations will be unable to compete with global platforms and tech giants for the talent they seek.

Redeployment may help companies play to workers' strengths, but skills gaps will inevitably remain. So in the AI era, companies should double down on retraining the people they have, with an emphasis on lifelong learning and adaptability.

Though reskilling (teaching employees new or qualitatively different skills) and upskilling (raising existing skill levels) have become hot topics, there seems to be more talk than large-scale action in these areas. A key choice is whether to use in-house training resources and programs tailored to the company or to partner with an educational institution to provide external learning opportunities for employees. AT&T has chosen the latter course: The company has developed a broad set of partnerships with 32 universities and multiple online education platforms to enable employees to earn the credentials needed for new digital roles.

Other companies, including the German software giant SAP and Walmart, have opted for in-house training programs. Walmart has set up more than 100 "academies" in the United States that provide classroom and hands-on training for frontline supervisors and managers. SAP has constructed a series of "learning journeys" for thousands of employees at its Digital Business Services division that feature boot camps, shadowing senior colleagues, peer coaching, and digital learning.

Both approaches make workforce learning a priority. Partnering with external educational providers may be easier for most companies for the same reason that most organizations can't afford to compete with deep-pocketed tech giants to hire experienced tech talent. A starting point in either case is to take stock of the skills that are already present in the workforce and then map those skills to the skills companies are projected to need in the future.

Taking inventory is also a first step on the way toward making HR more strategic, as many companies will need to do in the next few years. With strategy likely to be highly dependent on the availability of talent in an AI-fueled future, HR will need to play a larger role in long-term planning, with the chief human resources officer evolving into as central a figure as the CFO.

HR departments will also have to undergo profound changes in the way they work. That means developing an internal market for talent as well as a marketplace for lifelong training experiments—and supporting employees' learning by analyzing skill building by career path, for instance, and focusing on closer human-machine "interwork." The long-term goal is to embed a new flexibility and adaptability in the workforce, accompanied by a new adaptability within the HR function.

Ensuring that the right skills are in place at the right time is shaping up as one of the biggest corporate challenges of our time. Given the winner-takes-most dynamic that we are already seeing, first with digital transformation more broadly defined and now in more pronounced fashion with AI and automation, no company can afford to underestimate the coming skill shifts and how those may affect their prospects.

5

Getting Your Employees Ready for Work in the Age of AI

Eva Sage-Gavin, Madhu Vazirani, and Francis Hintermann

The era of AI is upending work as we know it. And as companies start using intelligent technologies in earnest, many people who have been well-trained for their positions for a long time may suddenly find themselves in uncharted waters.

The good news is that employees are ready to embrace the changes they see coming. According to an Accenture survey on the future workforce, over 60% of workers have a positive view of the impact of AI on their work. And two-thirds acknowledge that they must develop their *own* skills to work with intelligent machines.

Large companies, however, are not on the same page as their employees. For one thing, business leaders believe that only about one-quarter of their workforce is prepared for AI adoption. Yet only 3% of business leaders are planning significant increases in their training budgets to meet the skills challenge posed by AI.

How can companies and employees find common ground when it comes to skill development and investment in AI capabilities? To start, senior executives should seek clarity around capability gaps and determine *which* skills their people need.

From there, leaders should take an approach that advances those skills for human-AI collaboration.

Understanding the Missing Middle

Much of the press around AI has focused on automation and what it will mean for jobs. While these predictions often run the gamut of completely devastating to highly positive, continued progress in technology has stoked fears in people that AI may someday make their jobs obsolete.

However, much of the leading data around jobs suggests humans will continue to play a major role in the AI workforce—although a transformed one. Using the US Department of Labor's O*Net database, we analyzed more than 100 abilities, skills, tasks, and working styles in the United States over the past decade and found a sharp rise in the importance of creativity, complex reasoning, and social and emotional intelligence—the uniquely human skills—in many jobs.

Employees will need to apply these skills when using the array of new technologies now appearing in the workplace. They will also need to learn how to use AI to augment those skills—to make the "human + machine" interaction add up to more than the sum of its parts. We call this interaction *the missing middle* because companies often focus entirely on the ends—on what people can do without AI-powered machines or on how machines can automate work previously done by people. But the greatest value comes from the two working together.

This new era of AI calls for a new approach in business. First, both companies and employees must show that they are mutually ready to adapt to a world of work built around people and intelligent machines; identifying new tasks and skills needed to perform them is critical. Second, educators and learners both within companies and in other institutions must embrace science and smart technologies to speed up learning, stretch thinking, and tap latent intelligences. And third, employers and workers must create and maximize the motivation to learn and adapt over their lifetimes.

Establish Mutual Readiness

Companies should commit to long-term investments in workforce skills development, while employees should start adapting their skills for an AI-enabled environment. However, this readiness to change is feasible within an organization only when both company and worker have opportunities to realize their common aspirations in the new workplace. One key to achieving this: Identify new tasks and the skills needed to perform them. Then map the company's existing internal capabilities to new roles, and identify where training and new skills will be necessary.

PlainsCapital Bank, one of the largest independent banks in Texas, reimagined the nature of the work its people were doing. After it introduced digital banking services, demand for human bank tellers started to decrease. The company then combined the tasks of onsite teller, adviser, and customer service agent, creating the role of *universal banker*. Bringing these disparate roles into one was not a simple change. The requirements for success in the new role include excellent interpersonal skills, strong problem-solving abilities, and creativity, in addition to knowledge of the products and the customer experience.

To fill these new jobs, the bank changed the selection process to behavioral-based interviewing, based on the belief that how someone has acted in past situations can be a predictor of future performance. Today, when PlainsCapital finds a good match from a pool of internal candidates and identifies where they need skills development, the organization provides the relevant training, whether in technical skills, socio-emotional skills, or both.

Accelerate Ability

Adapting to AI is not just a matter of more training. It's also about *new* training. How new? Consider what neuroscience tells us.

Neuroscience is providing evidence that both cognitive and noncognitive skills can be taught to people of any age. Brain plasticity shows that under the right learning conditions, the adult brain can restructure itself in remarkable ways. Companies need to take this evidence to heart to stay nimble in the age of AI.

Companies will increasingly need employees to collaborate effectively with coworkers from different disciplines. Mental-training approaches such as "mindfulness"—teaching people to be aware in the moment—are not traditionally part of corporate training. But they have proven effects in strengthening employee performance. For example, employees who took part in Accenture's mindfulness program (designed with mindfulness-training company Potential Project) reported notable improvements in their ability to focus, set priorities, and collaborate within teams.

Another way to accelerate ability within organizations includes using virtual training. Instead of having to go to a classroom, an employee can learn online from wherever is convenient for them, on their own schedule. AI-based adaptive-learning systems guide employees through computer-based courses, monitoring their progress, personalizing lessons, coaching, and providing feedback.

Education companies are bringing the best of AI and neuroscience to corporate learning and development. Startups like the Silicon Valley–based Socos Labs, the Lausanne, Switzerland–based Coorpacademy and the Paris-based InsideBoard offer adaptive learning experiences using AI algorithms and scientific principles. Accenture's Future Talent Platform uses virtual reality and augmented reality to simulate real-world situations in training; employees make decisions given the information they see, and then receive feedback in real time to develop their socio-emotional skills.

People often learn best from each other. At Google, workers learn new skills on the job. Some 80% of all tracked training at Google is now done through the g2g (Googler-to-Googler) voluntary network of 6,000-plus employees. Companies may also encourage skill-building through "outside-in" talent exchanges with startups, universities, NGOs, and the public sector.

Solidify Shared Values

What values matter most in the evolving world of AI and humans at work? At the highest level, it's about sharing the commitment to education amid rapid change. People need time to adapt and prepare for new forms of work, and companies must recognize individual needs. Employees need the freedom to pursue skills development that is aligned to their passion and purpose at work; companies should subsidize training programs with external stakeholders. Success in the future of work means adaptability to constant change, and therefore lifelong learning is crucial for workers.

Lifelong learning can open doors to new careers at any stage. Accenture has been using its Future Talent Platform to train more than 165,000 people globally in the latest digital technologies for the past two years. Users can explore more than 3,500 learning boards curated by emerging technology experts within the company and from its partners to develop skills in critical areas such as digital, cloud, security, and artificial intelligence.

Similarly, Salesforce's learning platform, Trailhead, helps employees acquire the skills they need to change roles. For instance, some have learned how to code on Trailhead and then moved from job recruiting or sales into engineering roles. By earning and displaying badges of achievement, employees showcase their transferrable skills; Salesforce benefits by keeping its large staff up to date on the latest technical skills.

A Public Matter

Embracing new technology requires a continual reimagining of work and ongoing skills development. Meeting the demand for skills training may require more resources than a company can muster. Here, public–private partnerships can and should be brought to bear to great effect.

Singapore is offering individual learning accounts to address the need for training. Any citizen over 25 can get a tax credit for taking training from any of 500 approved providers. In 2017, more than 285,000 Singaporeans used the credit. And in the United States, the Aspen Institute Future of Work Initiative has proposed tax-advantaged lifelong learning and training accounts, which would be jointly funded by employers, employees, and government.

Our research has found that nearly half of employees believe their ability to develop new skills is impeded by lack of time and sponsorship for training from their employers. It's time to show them their company has their back.

6

The Challenge of Scaling Soft Skills

Lynda Gratton

It is becoming increasingly clear that for most working people, a proportion of the working tasks they currently perform will be either completely replaced by machines (AI if the tasks are cognitive, robots if they are manual) or augmented by a human-machine interface.

While there is less clarity about the types of tasks that will remain within the human domain, we can make some predictions. We know that, right now and in the foreseeable future, machines are generally poor at understanding a person's mood, at sensing the situation around them, and at developing trusting relationships. So as the World Economic Forum report on future skills argued, it is human soft skills that will become increasingly valuable—skills such as empathy, context sensing, collaboration, and creative thinking.

That means that millions of people across the world will have to make the transition toward becoming a great deal better versed in these soft skills.

But that's far from easy. The paradox is that while we understand a lot about how to develop the "hard skills" of analysis, decision-making, and analytical judgment, we know a great deal less about the genesis of soft skills. Perhaps more important, much of the context of how people learn and perform is currently skewed toward hard skills. Understanding the obstacles to developing soft skills and then addressing them is crucial for our schools, our homes, and our workplaces.

Three Barriers to Developing Soft Skills

It seems to me that there are three major barriers to scaling the development of soft skills.

Schools are too much like factories. The basic foundations for most schooling systems were laid down after the Industrial Revolution. The aim by the early 1900s was clear: to take a population that was mainly engaged in craft or agricultural work and prepare it for work in factories—and, more recently, offices. Though some schools have moved the curriculum to soft skills and creativity, in many schools, these traditions hold firm. Children are trained to stay still for hours at a time (as they would on a factory production line), to engage in rote learning, and to be compliant and follow rules. The pity of this is that these skills are ones at which machines are highly competent. More important, these conditions do little to nurture in children the skills of compassion, inventiveness, and being able to interpret people correctly.

The home is saturated with technology. There is mounting evidence that technology use is affecting the development of human soft skills. When children and adults spend a significant amount of their time engaged with virtual online games and social media, for instance, there is some evidence that their face-to-face social skills begin to atrophy. Short volleys of social interaction do little to support social skills. This is important, because the evolutionary benefits that humans have developed in empathy and collaboration need to be reinforced in subtle individual learning. Contrast, for example, a child's conversations with the Amazon.com virtual assistant, Alexa, and with an actual adult. In interacting with Alexa, the child may be tempted to bark instructions and possibly be rude to the machine. Alexa simply replies in a steady, dignified manner. If a child mimicked such an interaction with an adult, he or she would likely be reprimanded for rude behavior.

That is not to take away from the fact that technology could play a significantly beneficial role in the development of soft skills. Over the last decade, there have been major developments in technology-based learning, including online programs that tens of thousands of people can participate in. The primary focus of the majority of these programs has been on helping people develop hard skills. These programs are very competent at teaching content, simulating decision-making, and testing for knowledge. But what these learning technologies have not yet cracked at scale is how to support the development of soft skills across thousands or millions of people. Those that do teach these skills tend to be small-scale initiatives that involve face time and mentoring.

Stressful work reduces empathy. Finally, there is the challenge of the context of work itself. There is clear evidence that most adults learn a great deal at work, so we could imagine that adults could learn soft skills in the workplace. But it turns out that the development and use of soft skills such as empathy and creativity are highly sensitive to how a person is feeling. Studies show that when people feel under pressure, like they're being treated unfairly, or otherwise under stress, the hippocampus—the part of the brain's limbic system that is associated with emotion—is much less able to engage in empathic listening or appreciating the context of a situation. The brain, in a sense, closes down to learning or performing soft skills.

The challenge is that many workplaces have practices and processes that, often unintentionally, result in high levels of stress. Moreover, the antidotes—such as more flexible working conditions, collaborative cultures, the institution of fair processes—are not adopted quickly.

A Way Forward

Historical accounts of the effect of the Industrial Revolution show that the introduction of new technologies in that period threw many people's lives into confusion as they struggled to reskill to meet the needs of new machines. This period, known as Engels' Pause, resulted in deep unhappiness and a reduction in productivity before people upskilled and society was redesigned.

It is clear that there is no easy answer to this challenge. But the need is great, and the speed of implementation is crucial. Looking around, there are some exciting projects and initiatives—but to really make a difference, these must be adapted more broadly. To avoid the same kind of turmoil today that was seen during the Industrial Revolution, we need to act now.

First, it's crucial that schools remove the textbooks and rote learning and instead focus on empathy and collaboration, as High Tech High in San Diego, California, has done.

Second, there's a need for businesses to focus more on technological innovation to help employees develop soft skills.

There are already some exciting initiatives that might provide useful models worthy of rolling out to larger numbers of people: Fast-food giant KFC uses a training program that emphasizes customer interaction, and Fidelity Investments is experimenting with using virtual reality to upskill its call center employees in empathy and customer insight.

And third, at work, technological innovations can help reduce stress. Right now, wearables capable of tracking heart rate, skin temperature, and brain waves can pinpoint sources of stress more accurately—and therefore help people monitor the "rested brain" that is most capable of being empathic. Designing work to engage the rested brain will include minimalizing technological distraction and creating quiet periods during the day.

Some of these solutions are truly scalable, particularly those that use cost-effective technologies. But we cannot wait on the sidelines for Engels' Pause. We must seize these ideas and invent others. If we are to retain our humanity in the age of machines, we need to bring to the fore what it is to be human.

7

The Time for Retraining Is Now

R. Edward Freeman and James R. Freeland

None of us know how our technological future will unfold. Just recently, we learned that Amazon Go will be opening more cashierless, no-salesperson stores and that a burger chain has "employed" a robot to flip its burgers. At the same time, Uber had put on hold its use of self-driving vehicles after a fatal accident in Arizona (although the company is now testing these vehicles again). The times are a-changing, and the robots are here. As we develop more sophisticated AI problem-solving approaches, we know that the employment landscape will be changing rapidly.

We also don't know whether the doom and gloom pundits are correct or not. Bain claimed in a February 2018 study that automation may eliminate 20% to 25% of jobs in the US by 2030. A UK study put the estimate much higher—it predicted that as many as 47% of current US jobs could be made redundant or irrelevant in a short time frame. And a November 2017 McKinsey Global Institute report suggested that by 2030, as many as 375 million workers around the world may need to switch occupational categories—that's a full 14% of the global workforce.

Others tell us that the situation is not so dire. New technology, they say, has always turned out to be a net job creator over time, despite predictions about job elimination.

But even if the more optimistic prospects are true, will the prosperity created by these new jobs be shared broadly? And what do we need to do to be prepared?

We Need to Prepare for Different Skill Sets

Whatever the net increase or decrease in jobs overall, what is not at issue is that these will be different jobs, requiring different skill sets.

Rather than wait to see the answer about whether more jobs will be lost or be gained, we need to act now to enable current employers and employees to gain the skills they are going to need in the brave new world of AI technology. Let's look at some examples of what is currently being done.

Technology companies are helping train young adults. IBM has designed Pathways in Technology (P-TECH) schools, partnering with almost 100 public high schools and community colleges to create a six-year program that serves large numbers of low-income students. The skills learned are reinforced through mentorships and internships with IBM and many other companies. Graduate rates are four times the average, and those getting jobs are at two times the median salary.

Technology companies are helping retrain existing employees. Bit Source, a software development company based in Pikeville, Kentucky, is training Kentucky coal miners to become software developers. The company's aim is to move workers from "exporting coal to exporting code." Wind technology company

Goldwind Americas, a subsidiary of Xinjiang Goldwind Science & Technology Co. based in Beijing, China, launched Goldwind Works to retrain employees from the coal, oil, and natural gas industries in Wyoming to become wind turbine technicians, a category expected to grow over 100% in the next few years. And Flexport, a San Francisco–based freight company, developed an initiative to train incoming workers to work in the digital age in shipping logistics, where it has been able to compete with much larger companies. Currently these programs are small, but they illustrate what can be done.

Technology companies are providing new and broader digital education to their own workers. Cognizant Technology Solutions Corp., an IT company with over 250,000 employees, recently retrained over 100,000 in new digital skills. Given the variety of digital skill sets, ongoing training is seen as a prerequisite for company success. The company says it sees retraining as going together with managing attrition and automating key processes. Similarly, when communications giant AT&T discovered that over 100,000 of its employees were in jobs that wouldn't be around in 10 years or less, it committed to a massive retraining effort. AT&T's Workforce 2020 uses both partnerships with traditional degree-granting universities such as Georgia Tech as well as course bundles of massive open online courses (MOOCs) to grant "nanodegrees" for courses, mostly online, that can be individualized with certifiable skills. In the words of one executive, "AT&T is working to instill a mindset in which each individual becomes CEO of his or her own career, empowered to seek out new skills, roles, and experiences."

Governments are getting in on the act, too. Tennessee made community colleges free for adults without a degree starting in the fall semester of 2018. Oregon has free community college for all high school graduates, including those with GEDs. San Francisco has made community colleges free for all of its residents. The State of New York has made both two- and four-year public colleges free for students with families who earn less than $125,000.

Other countries are moving in the same direction. A World Economic Forum white paper presents examples from Singapore, Denmark, the UK, and Brazil of how governments in other countries are getting involved in providing resources for retraining and lifelong learning.

Four Suggestions for Learning in the AI-Powered Future

There are many more examples of how companies and governments are tackling this problem. But we need more.

We want to make four suggestions for retraining and lifelong learning as we move into this new digital, AI-powered age.

First, we need to see the role of governments, at least partially, as facilitating value creation. Traditionally, government has been seen as referee, regulator, and redistributor. These are important roles. However, helping businesses, both for-profit and nonprofit, facilitate the value creation they do so well is also a legitimate role. The government examples above are good illustrations of that point. Facilitating lifelong learning helps companies find employees who can create value for their stakeholders and improve the economy and society.

Second, we need to become a nation, and indeed a world, of entrepreneurs. The trend of fewer startups and fewer workers in startups must be reversed. We need entrepreneurship that will create more opportunities for new businesses. We need to ask what governments can do to facilitate the entrepreneurial spirit. We need a massive effort to help people who have been shut out of the value creation system to get the skills and the capital they need to start and grow businesses. We believe that this effort needs to be focused locally and started in elementary schools where we teach kids that creating value for others is what business is really about. And we need to encourage their ideas, even at that age, for how to do it.

Third, we need to become a nation of experimenters. "One size fits all" programs generally do not work for many people. We need to try lots of things that will provide new marketable skills and mindsets for learning, figure out how to keep what works, spread the news, and stop doing those things that don't work. We need experiments that use evidence-based results to understand what is scalable. The Opportunity Insights, now directed by Harvard University, is an excellent example of the kind of research that can make a difference. AT&T's approach won't work for all businesses, and Denmark's system won't work in all countries. The locus of action must be local, even though we will need organizations with a more national and global focus to spread the good ideas.

Finally, we need to encourage lifelong learning, especially for those at the lowest rung of the economic ladder. Many people are far from using the resources of MOOCs, the web, and other available technologies. Why not open our public schools to lifelong learning, where courses could be taught in the evening, and everyone could get certifiable skills? Local companies could contribute as part of their mission to be community builders. Local governments and companies could help with any funding that would be necessary. Many citizens could become teachers and mentors.

We have only scratched the surface of what needs to be done. There is an urgency to get started now and build on the set of experiments that are currently going on. We must engage in this effort.

We want to be the ones in charge of our own future—let's not leave it to the robots and AI.

II

Strategy and Operations

8

Don't Let Artificial Intelligence Supercharge Bad Processes

Sam Ransbotham

Scenarios describing the potential for artificial intelligence seem to gravitate toward hyperbole. In wonderful scenarios, AI enables nirvanas of instant optimal processes and prescient humanoids. In doomsday scenarios, algorithms go rogue and humans are superfluous, at best, and, at worst, subservient to the new silicon masters.

However, both of these scenarios require a sophistication that, at least right now, seems far away. Our recent research indicates that most organizations are still in the early stages of AI implementation and nowhere near either of these outcomes.

A more imminent reality is that AI is agnostic and can benefit both good and bad processes. As such, a less dramatic but perhaps more insidious risk than the doomsday scenario is that AI gives new life to clunky or otherwise poorly conceived processes.

Consider faxes in the health care industry. Despite being obsolete in most places, faxes are still a fundamental part of the medical infrastructure. Because of their long history and strong network effects, the medical industry still sends a staggering number of faxes every day.

Invented in the 1840s, well before the telephone, faxes illustrate how difficult it is to change an entrenched process.

Despite widespread use, sending faxes is, for the most part, a horrific way to transfer information. The process is typically (1) extracting and printing information from a computer system, (2) scanning it into an image, and (3) transmitting it via fax. The burden rests squarely on the recipient to interpret a pixelated approximation of the original information. Structured digital information has become unstructured. With every fax, a data scientist gets his or her wings ripped off.

The conversion from structured information to unstructured and back is a waste. No one wins—a patient may be waiting for approval, medical staff may lack information, errors can creep in, and so on. At a minimum, time and effort are needlessly spent.

Advances in AI and image processing are making significant progress in reducing this problem. Organizations can use AI to recognize images, automate the interpretation, and restructure the information. Certainly, this is a welcome improvement. No one gets competitive advantage or business value from wasted time spent restructuring information.

However, this improved image processing is a pyrrhic victory. While it may be more efficient, perhaps even significantly, the expended resources are lost forever. In the absolute best case, the original structured information is recovered, but it will be expensive and difficult to get close to that best case. The realistic case is far less promising.

The danger is when AI gets just good enough to let a less-than-ideal system like this limp on. Improved image processing can allow organizations to cut costs and automate much of the processing (this applies beyond faxes and beyond health care), but AI may also mask the symptoms of bad processes. It provides a bandage, not a cure.

If AI were to fail completely at interpreting faxes, we might be better off: The system would be untenable. Costs would rise. People would notice. Change would be inevitable.

Yes, it is better to reduce this annoying work than to continue doing it. This is a justifiable reason to apply AI in organizations and a great example of reducing the scut work. However, gains in AI applied to bad processes may staunch wounds, not heal the organization.

Successful organizations will continue to improve underlying processes. But, in many industries, like health care, with legacy systems and embedded processes that involve many people and many organizations, that will be difficult.

One risk we see is that upstart organizations, unencumbered by legacy processes, will be able to start fresh. They won't have to make significant investments of time and resources to apply AI to processes that perhaps should not exist in the first place. These new entrants, possibly from outside your industry, will apply other tools and technology to completely bypass the process your organization is struggling to bandage with AI.

Are there processes in your organization that AI can improve? Great. But, be sure to ask yourself: Should these processes exist in the first place? No amount of gee-whiz AI used to improve a process beats not having to rely on that process at all.

9

Unlocking the Value of Augmented Reality Data

Joe Biron and Jonathan Lang

The word "sensor" has become inseparable from the internet of things (IoT), where sensors detect environmental conditions and communicate these signals bidirectionally as data, whether it's an industrial machine reporting its operating condition or your home thermostat being turned on remotely. This data is a key driver of the IoT's global economic impact, which McKinsey estimates could reach up to $11.1 trillion per year by 2025.

Before the IoT, the two key functions of a sensor—to detect environmental change and to communicate that data—were largely carried out by humans. Today, as augmented reality (AR) technology gains adoption, humans will soon be equipped with sensors through various AR devices, such as phones and headsets. This augmentation provides uncharted opportunities for organizations to use these data insights to drive operational effectiveness and differentiate their products and services for consumers.

The AR market today is similar to where the IoT market was in 2010, generating considerable buzz and proving early value from new capabilities for users. AR's capacity to visualize, instruct, and interact can transform the way we work with data. Based on the lessons learned in the early days of the IoT, enterprises should be asking the question: What's the best way to plan for AR device data and see its value, so we can build better products and processes from user insights?

Smart, Connected Reality Means More User Data

As we do that planning, there is much to learn from our recent, connected past. The 2007–2008 iPhone and Android market releases provided significant data about how customers engaged with their brand, and it gave engineers new insights into user requirements. This market disruption flipped the value proposition—applications could sense and measure customer experience in conjunction with delivering it, and it opened the door to subscription- and use-based services. With similar sensing capabilities emerging through the IoT for physical products, companies quickly built in connectivity, giving rise to smart, connected products (SCPs) that make up the internet of things. The data arms race and the emergence of the data economy has been disrupting technology laggards ever since.

Considering these proven market dynamics, the potential for AR-as-a-sensor being the next-generation modality for gathering rich data is profound. Products are already equipped with APIs and connectivity, and AR devices are loaded with machine sensors, from multiple cameras to GPS, Bluetooth, infrared, and accelerometers.

AR also unlocks a set of sensors that are often forgotten amid the frenzy of machine automation—the human capacity for creativity, intuition, and experience. Humans have incredible abilities to recognize and react to novel situations, assessing them more quickly and accurately than current technology systems can.

Consider what humans using AR devices could add to such interactions. There are valuable new data and behavior insights to gather from both unconnected products and SCPs. For an unconnected product, a person using AR-as-a-sensor tech might ask: How is this product used, and what are the user preferences? For an SCP, he or she might ask: How does usage affect performance, and how can this product adapt to usage?

Human interaction offers context to answer how both unconnected and connected products are really being used, how they perform, and how they can be adapted for their purpose by bringing human creativity, intuition, and flexibility to AR data gathering.

Assessing the Business Opportunities from New AR Data

The new data created by AR establishes a feedback loop to answer questions about how a product is being used or where opportunities for customization exist. The value of this type of customer data has increasingly become core to business strategy in the new digital economy.

One way to understand the value of your data is to assess it using the DIKW Pyramid, a hierarchy used in information management for understanding the transformation of raw data signals into value-rich knowledge and insights.

If this looks familiar, it's because this is exactly the type of data flow that creates value for manufacturers and users of SCPs today. By feeding these insights back into their engineering

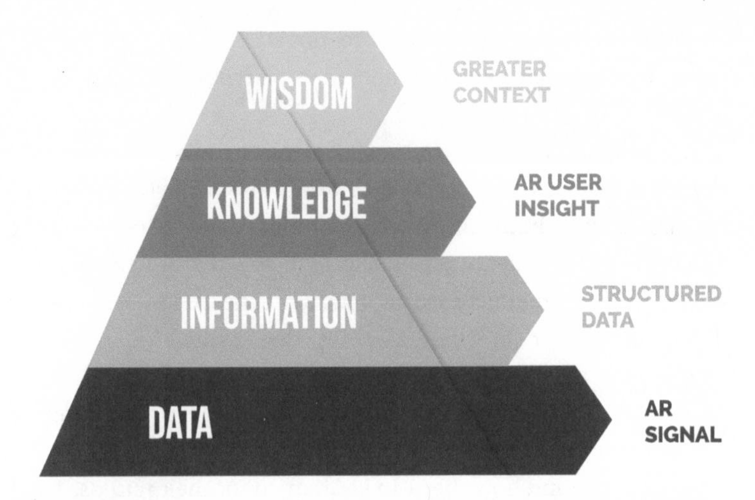

The DIKW model The DIKW model shows the hierarchy of raw data and value-rich wisdom from its analysis.

systems, companies can optimize their product portfolio, design, and features like never before.

AR-driven data collection can be combined with IoT data streaming from SCPs to drive additional context and generate more complete insights. For unconnected physical products or digital-only services, humans, interacting with these via AR, can act as sensors to deliver new insights about product or service use, quality, and ultimately about how to optimize user experience and value.

Early Use Cases for AR Insights

There are a wide range of early example use cases for these kinds of data sources. Companies like Honeywell, Cannondale, Amazon.com, and DHL have created new opportunities for product strategy, value chain, and quality control activities by utilizing

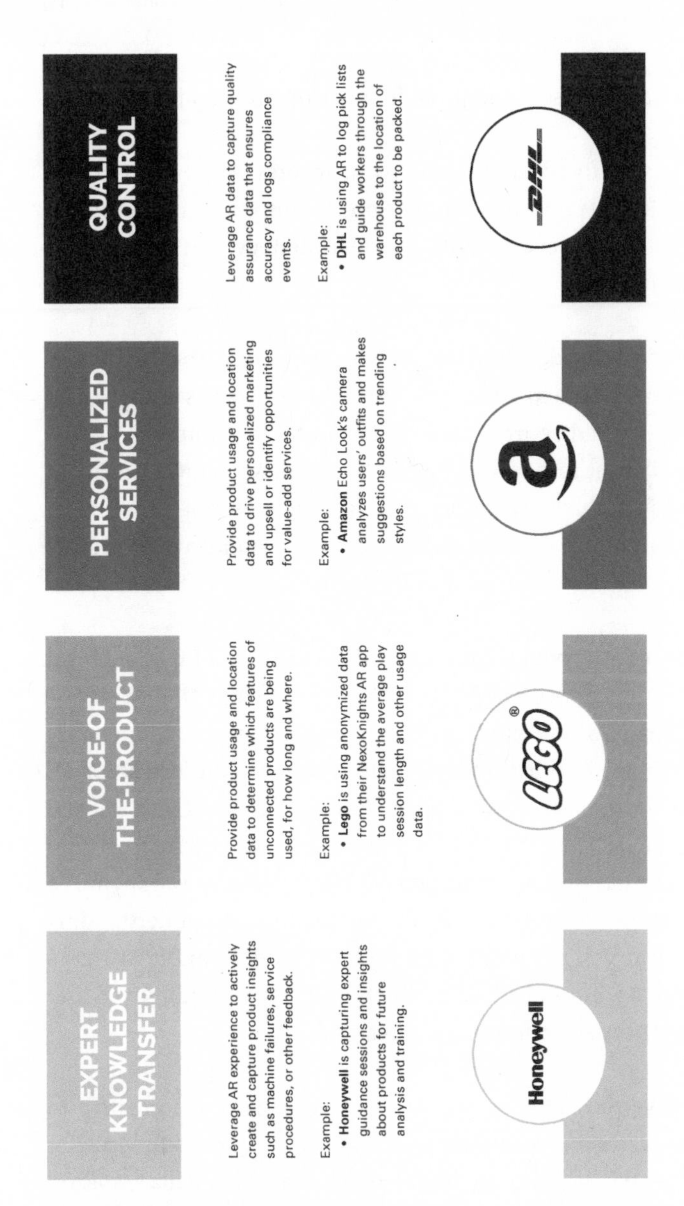

Early AR Use Cases Companies are already using insights from AR data to create new products and services.

AR data from users and providing personalization based on this data.

These early examples clue us in to how AR-as-a-sensor will make its way into the mainstream, creating new opportunities for manufacturers, in both product strategy and value chain activities.

Expert knowledge transfer. While many tout the benefits of delivering AR experiences to users, Honeywell is using AR to capture expertise from seasoned workers and improve knowledge transfer to new employees. By digitizing knowledge about a product that is revealed only through experience, Honeywell can understand products and their use in ways previously unavailable, without using embedded sensors.

Voice of the product. For Cannondale, its newest high-end bikes are being shipped with accompanying AR apps that showcase bike features and guide users through common maintenance procedures. This is fundamentally changing the definition of the product from physical bikes alone to combined physical and digital experiences. Thanks to these digital AR experiences, Cannondale has the opportunity to gather and analyze anonymized data to deliver the "voice" of the product. By understanding what features and procedures are most used, Cannondale has a potential window into its products that can drive improved user experiences and competitive advantage.

Personalized services. AR is billed as being transformative to e-commerce and retail, because it allows customers to visualize and try before you buy, unlike other available media. Amazon Echo Look is a new device allowing customers to capture and see virtual clothing on themselves before purchasing the real thing. In January 2018, Amazon patented "magic mirror" technology, which combined with the Echo Look, will pave the way for the next-generation dressing room. The data captured today through the Echo Look is being analyzed to create user preference profiles and curate suggested purchases based on tastes. It isn't hard to imagine how, combined with the ability to augment those clothing suggestions back onto the customer, this new source of AR data will lead to a new level of personalized services and experiences.

Quality control. DHL has long been at the forefront of AR technology and is on the advanced end of current AR programs. By reducing friction across logistics processes, AR delivers great value for DHL's employees as they go about daily tasks. But this data does not end with the user. Using integrated computer vision to perform the task of bar code scanning, DHL now has a way to capture and log quality assurance data, allowing the company to understand where human behavior may affect order quality and process efficiency.

All of these companies' early implementations give a glimpse of what is to come in AR experience delivery and how that data can create additional value for businesses.

Connecting the Strategic Dots

What about the impact to a broader data strategy? Taking a step back to this level, the implications are potentially significant. The value of many data initiatives hinges on the ability to connect the dots. While IoT, digital engagement, voice of the customer, and other initiatives continue to create significant opportunities to optimize products and processes, many enterprises are running these projects in silos because of technological or organizational constraints.

As AR emerges as a new source of context-rich data, companies that connect the dots between multiple sources from smart, connected products to CRM data, digital engagement, and other sources of insight will create the greatest opportunities.

Enterprises that want to capitalize on these opportunities should create cross-functional leads or tiger teams dedicated to the desired outcome—improving the customer experience—rather than by the traditional functional or technology-oriented alignments.

In this new data-driven world, the whole is greater than the sum of the parts, and AR just might be the missing piece you need to complete your vision.

10

Unpacking the AI–Productivity Paradox

Erik Brynjolfsson, Daniel Rock, and Chad Syverson

We see the effects of transformative new technologies everywhere except in productivity statistics. Systems using artificial intelligence increasingly match or surpass human-level performance, driving great expectations and soaring stock prices. Yet measured productivity growth has declined by half over the past decade, and real income has stagnated since the late 1990s for the majority of Americans.

What can explain this paradox?

Our close examination of recent patterns in aggregate productivity growth highlights the apparent contradictions. Examples of potentially transformative new technologies that could greatly increase productivity and economic welfare abound, as noted in the 2014 book *The Second Machine Age*. For instance, consider the recent progress in areas such as machine image recognition. At the same time, productivity growth has been historically slow over the past decade.

And the sluggishness is widespread, occurring not only in the United States but also in other nations of the Organisation for

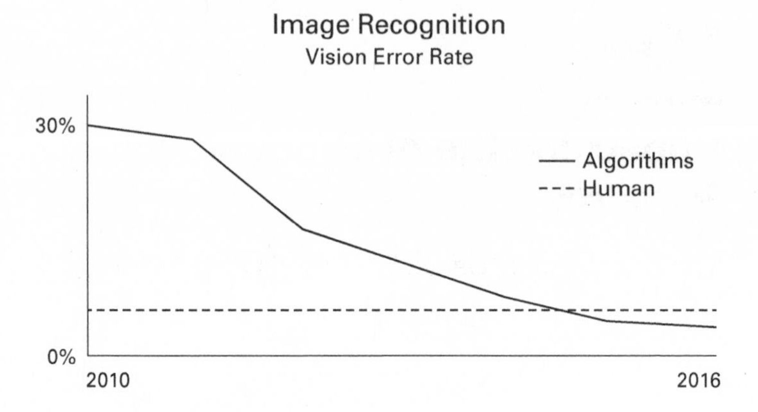

AI versus Human Image Recognition Error Rates

Economic Co-operation and Development (OECD), as well as among many large, emerging economies.

Some pessimism about future technological progress has spilled over into long-range policy planning and corporate strategy plans. The US Congressional Budget Office, for instance, reduced its 10-year forecast for average annual labor productivity growth from 1.8% in 2016 to 1.5% in 2017. Although modest, that drop implies US gross domestic product will be considerably smaller 10 years from now than it would if productivity simply continued to grow at the same rate as before—a difference equivalent to almost $600 billion in 2017.

Nevertheless, when we review the evidence, we come to a different conclusion and take a more optimistic view. The recent productivity slowdown says nothing about future productivity growth and is no reason to downgrade prospects. In fact, history teaches the opposite lesson. Past surges in productivity were driven by general-purpose technologies (GPTs) like electricity and the internal combustion engine. In turn, these technologies required numerous complementary co-inventions like factory redesigns, interstate highways, new business processes, and changing workforce skills before they truly fulfilled their potential. Importantly, these co-inventions took years or even decades to materialize, and only then did productivity improve significantly.

We believe that AI has the potential to be the GPT of our era. And like earlier technologies, it requires numerous complementary innovations—including new products, services, workflow processes, and even business models—that are often costly and time-consuming to develop. The low productivity growth of recent years may partially reflect these costs and may also be a harbinger of significantly higher growth once necessary co-inventions are put in place.

Accordingly, we see no inherent inconsistency between forward-looking technological optimism and backward-looking disappointment. Both can simultaneously exist. Indeed, there are good conceptual reasons to expect them to simultaneously exist when the economy undergoes the kind of restructuring associated with transformative technologies. Future company wealth and historical economic performance show the greatest disagreement precisely during times of technological change. Our evidence suggests that the economy is in such a period now.

Four Explanations for the Paradox

Specifically, we found four possible reasons for the clash between expectations and statistics: (1) false hopes, (2) mismeasurement, (3) concentrated distribution of gains, and (4) implementation lags. While a case can be made for each of these four explanations, implementation lags are probably the biggest contributor to the paradox. In particular, the most impressive capabilities of AI—those based on machine learning and deep neural networks—have not yet diffused widely.

The first three reasons rely on explaining away the discordance between high hopes and disappointing statistical realities rather than accepting both of these sets of claims. In the misplaced optimism scenario, it's the expectations of technologists and investors that are off base. The mismeasurement explanation means that it is the tools we use to gauge reality that are inaccurate. And in the concentrated distribution theories, private gains for the few don't translate into broader gains for the many. The technological promise is there, but dissipative activity prevents the technological returns from reaching most people.

The fourth explanation allows both halves of the seeming paradox to be correct: There is good reason to be optimistic about the productivity potential of new technologies while also recognizing the recent slowdown. It takes considerable time—more than is commonly appreciated—to sufficiently harness new technologies.

There are numerous cases where we see a lag between tech achievements and economic impact. Retailers' recent experience with e-commerce is a good example. The e-commerce excitement of the 1990s was prophetic, but it took nearly two decades—until 2017—for online business models to approach 10% of total retail sales. The sector as a whole required the build-out of an entire distribution infrastructure. Customers had to be "retrained" to buy online. Organizational inertia held back innovation in business processes, supply chains, and product selection. None of the needed changes happened overnight, even though the potential of e-commerce to revolutionize retailing was widely recognized, and even hyped. The actual share of online commerce was a miniscule 0.2% of all retail sales in 1999. Only now are companies like Amazon.com having a first-order effect on more traditional retailers' sales and stock market valuations. Self-driving cars, medical applications of machine learning, and many other AI breakthroughs will likely follow a similar trajectory.

AI Ramping Up to Its Full Potential

As a GPT, AI will ultimately have an important effect on the economy and public welfare. At the same time, profound and far-reaching restructuring requirements will continue to prolong how long it takes to see the full impact on the economy and society.

What's more, and what business leaders may find most relevant, is that the required adjustment costs, organizational changes, and new skills can be modeled as intangible capital. A portion of it is already reflected in the market value of companies. However, going forward, national statistics will need to be reinvented to measure the full benefits of the new technologies and their true value.

Realizing the payoffs of AI is far from automatic and will require more fundamental changes than many executives typically imagine. We predict that the winners will be those with the lowest adjustment costs and the right complements. Companies that can best adapt to the new environment will find great opportunities, while competitive pressure awaits those that do not respond nimbly enough. With historical perspective and a realistic road map, we will all be prepared to share in the eventual benefits.

11

Using AI to Enhance Business Operations

Monideepa Tarafdar, Cynthia M. Beath, and Jeanne W. Ross

Artificial intelligence invariably conjures up visions of self-driving vehicles, obliging personal assistants, and intelligent robots. But AI's effect on how companies operate is no less transformational than its impact on such products.

Enterprise cognitive computing—the use of AI to enhance business operations—involves embedding algorithms into applications that support organizational processes.[1] ECC applications can automate repetitive, formulaic tasks and, in doing so, deliver orders-of-magnitude improvements in the speed of information analysis and in the reliability and accuracy of outputs. For example, ECC call center applications can answer customer calls within 5 seconds on a 24/7/365 basis, accurately address their issues on the first call 90% of the time, and transfer complex issues to employees, with less than half of the customers knowing that they are interacting with a machine.[2] The power of ECC applications stems from their ability to reduce search time and process more data to inform decisions. That's how they enhance productivity and free employees to perform higher-level work—specifically, work that requires human adaptability and creativity. Ultimately, ECC applications can enhance operational excellence, customer satisfaction, and employee experience.[3]

ECC applications come in many flavors. For instance, in addition to call center applications, they include banking applications for processing loan requests and identifying potential fraud, legal applications for identifying relevant case precedents, investment applications for developing buy/sell predictions and recommendations, manufacturing applications for scheduling equipment maintenance, and pharmaceutical R&D applications for predicting the success of drugs under development.

Not surprisingly, most business and technology leaders are optimistic about ECC's value-creating potential. In a 2017 survey of 3,000 senior executives across industries, company sizes, and countries, 63% said that ECC applications would have a large effect on their organization's offerings within five years.[4] However, the actual rate of adoption is low, and benefits have proved elusive for most organizations. In 2017, when we conducted our own survey of senior executives at 106 companies, half of the respondents reported that their company had no ECC applications in place. Moreover, only half of the respondents whose companies had applications believed they had produced measurable business outcomes. Other studies report similar results.[5]

This suggests that generating value from ECC applications is not easy—and that reality has caught many business leaders off guard. Indeed, we found that some of the excitement around ECC resulted from unrealistic expectations about the powers of "intelligent machines." In addition, we observed that many companies that hoped to benefit from ECC but failed to do so had not developed the necessary organizational capabilities. To help address that problem, we undertook a program of research aimed at identifying the foundations of ECC competence. We found five capabilities and four practices that companies need to splice the ECC gene into their organization's DNA.

Five Crucial Capabilities

We found that companies that successfully create value (that is, radically improve business processes to reduce costs and/or generate new revenues) using ECC applications possess five capabilities: data science competence, business domain proficiency, enterprise architecture expertise, an operational IT backbone, and digital inquisitiveness.

Data science competence. Data science competence encompasses a wide range of skills essential to ECC. It involves ensuring the availability and usefulness of massive amounts of data: collecting, cleaning, curating, tagging, and analyzing internal and external data from multiple sources. Such competence also entails identifying and describing relationships between data, as well as developing AI algorithms that have learned from data how to identify patterns and probabilities.

Top-notch data scientists have extensive knowledge in areas such as natural language processing, statistical inference, knowledge representation, and learning algorithms. Wipro, the Indian IT services company, includes these areas among the pillars of its data science expertise. Its data scientists deploy their skills and a variety of tools to create AI algorithms that can be inserted into enterprise applications.

For organizations that cannot develop the talent internally, obtaining data science competence is expensive and can require multiple hires from, for example, software development companies, technology consulting companies, AI startups, or university graduate programs in related fields. At a financial services company we studied—we call it OneBankAssure—the CEO hired a new direct report who was a technically accomplished data science academic and consultant. This person, in turn, hired the 20 data scientists who became the core ECC development team. Companies that are serious about ECC spend the money to hire the right data science talent. To raise the money, one pharmaceutical company we studied reduced its operational IT costs (by eliminating duplication in systems and standardizing processes across its business units) and redirected the savings to the acquisition of data science skills.

Business domain proficiency. Domain proficiency is needed to understand the tasks, workflows, and logic of existing business processes, as well as to imagine how ECC applications could improve them. As many organizations have learned the hard way, it's possible—even easy—to develop an elegant AI algorithm that uses massive amounts of data to learn how to predict or categorize something but doesn't improve the business. Having the right technical skills isn't enough. Domain proficiency links data science competence to business value.

For example, the ability of data scientists to effectively curate, tag, and analyze data depends on a clear understanding of the relationships among the data from a process and business point of view. Domain proficiency provides clarity around those relationships, referred to as ontologies. Data ontologies can become quite complex and even counterintuitive. Here's how a domain expert at a pharmaceutical company described some of the complexities he encountered capturing the data ontologies needed to support the company's research on diabetes: "A big part of diabetes is being overweight. Should there be an obesity dimension in our ontology of diabetes? Or is diabetes an attribute of obesity? Oh, and people who are overweight often have joint replacement issues. If they're overweight and their joints hurt and they have diabetes, the incidence of depression is very high, and dealing with depression is an important part of generating outcomes. Do I train the algorithm on depression?"

Domain proficiency is also important for creating the business rules that shape how the outputs from the algorithm are handled by the ECC application. For example, an ECC application that helps banks predict which customers are most likely to repay loans on time must include business rules for how the algorithm's prediction will be applied, such as: Will some loans be granted automatically? If so, under what conditions? With whom will the predictions be shared? Under what circumstances can a prediction be overridden?

For any given ECC application, domain proficiency is needed in all the functional areas that have a bearing on—or are stakeholders in—the operations of the focal process. For example, a team at a US bank that developed an ECC application to detect financial fraud needed proficiency not only in fraud identification and prevention but also in the related areas of regulatory compliance and banking law.

People with domain proficiency have deep process knowledge. They may be process owners, although they are often people with a regular hands-on role. Some companies seek to hire data scientists with domain expertise. Indeed, such individuals can partner well with business domain experts, but they cannot substitute for them when an ECC application is being developed. That's because they usually lack enterprise-specific knowledge about processes, policies, and practices currently in play.

Enterprise architecture expertise. Implementations of enterprise systems have a history of disappointing leaders who underestimated the organizational changes needed to capture their value. Too many leaders are reliving this disappointment with ECC applications. ECC applications do not deliver value by simply processing data and delivering outputs. They deliver value when the organization changes its behavior—that is, when it changes processes, policies, and practices—to gain and apply the insights from those outputs. Experts in enterprise architecture design the new organization needed to create business value from ECC applications, and they help manage the transition from the old organization to the new one.

The most ambitious ECC applications usually affect several, often fundamentally different business processes. In such cases, enterprise architects are needed to orchestrate the redesign of the systems, processes, and roles across organizational units. The more ambitious the ECC application, the more likely it will require far-reaching organizational changes.

Organization design and change issues can surface for seemingly small-scale applications. One medical drug distributor failed to recoup its investment in an ECC application that could accurately determine whether an online customer's insurance would cover a claim 90% of the time because the accounts payable department balked at making costly process changes required to support the application. If an enterprise architect had been engaged in the project from the outset, this loss might have been avoided.

The organizational changes needed to unlock the potential of an ECC application can be complex and intertwined. Enterprise architects are familiar with the organizational roadblocks that drive up costs or limit impact. At Wipro, enterprise architecture expertise helped smooth the way for a new help desk ECC application by first merging the company's existing help desk applications, reducing the types of fault tickets from 3,000 to 2,200, and eliminating redundancies in support tasks. By simplifying and standardizing the help desk process prior to the development of the ECC application, the company reduced and simplified the work of getting the data needed to train the AI algorithm, developing it, and ultimately automating the process, thus unlocking additional value.

Enterprise architects also recognize when ECC applications require changes in employees' jobs. They may see the need for upskilling, re-skilling, or the creation of entirely new roles. When a seemingly simple sales-lead-generating ECC application required its agents to do more cold calling and make more targeted pitches, OneBankAssure's enterprise architects designed a new coaching role to help agents that proved essential to generating benefits from the application.

Given the breadth of skills that enterprise architects draw on, this expertise can be difficult to develop. It often resides in people who are steeped in organization design and change management such as business leaders with experience managing technology-driven transformations or other reorganizations. Human resource professionals with exposure to a broad range of organizational roles can be a good source of architectural expertise in role design and redesign, as well as skills training. IT professionals with exposure to many different business processes, who can help streamline processes and establish the proper division of work between ECC applications and employees, also can be tapped.

Operational IT backbone. A company's existing technology and data foundation—its operational IT backbone—and the people responsible for it support the development and running of ECC applications. They supply the IT capabilities needed to store and access critical data, integrate ECC applications with other applications, provide reliable operations, and ensure privacy and security.

As noted earlier, for an AI algorithm to learn from data, a company must make available massive amounts of high-quality data that is cleaned and tagged. The lack of high-quality data is the most pernicious and least anticipated obstacle in the development of AI algorithms. OneBankAssure overcame this obstacle and accelerated its adoption of ECC by separating responsibilities for developing AI algorithms from responsibilities for providing the data. Since the IT unit already maintained the underlying operational infrastructure and good-quality operational data, it was able to support the algorithm developers by providing them with access to a data lake containing operational and external data. Responsibility for structuring the data for developing algorithms rested with the data scientists.

Almost no new enterprise application can operate in isolation from other enterprise applications. ECC is no exception. If an application is not properly integrated, it will be hard to use and possibly ignored. That's why the IT unit at OneBankAssure embedded the company's new sales lead system into its customer relationship manager system, which was part of its operational IT backbone. The CRM linked up-to-date contact information and customer history data to the sales leads. It also provided

Often, users of ECC applications must apply human judgment to predictions made by algorithms. They need to possess digital inquisitiveness – the inclination to question and evaluate the data before them.

a set of processes within which the ECC sales leads could be seamlessly presented to users. Being part of the IT backbone also meant that the sales lead system would be scalable, reliable, and secure.

The existing IT staff is the logical source for operational IT backbone expertise. At OneBankAssure, the IT function set the standards for ECC plug-ins and adapted applications to the company's production environment by refactoring and retesting the code. It also managed disaster recovery and security for installed ECC applications.

Digital inquisitiveness. The AI algorithms in ECC applications do not produce definitive answers. Rather, they produce predictions based on probabilities: the probability that a customer will buy a product, that a patient has a disease, that a loan will be repaid. Often, application users must consider these predictions and apply human judgment to arrive at decisions about how and where to promote offerings, what treatments to prescribe, or what loans to approve. To do this effectively, they need to possess digital inquisitiveness—a habitual inclination to question and evaluate the data before them. They must use that skill to better understand the options provided by ECC applications and continually improve outcomes.

The development of this capability requires a broad-based effort. A number of companies we studied instituted mechanisms to cultivate digital inquisitiveness. OneBankAssure's corporate university delivered a training program that introduced executives to the idea of using data effectively in making decisions. One exercise incorporated a strategy game, in which participants vied to develop the highest-value ECC solution to a business problem. At different phases of the game, they had to deal with poor-quality data (in fact, real company data), build decision trees, teach an algorithm to detect patterns, and develop a model to solve a problem. Wipro created an e-learning platform on which employees were able to take courses to understand what AI was, how it could be used in business processes, and how to work effectively in ECC-enabled processes. The company also trained hundreds of domain experts to act as AI champions throughout the organization.

Four Key Practices

Developing the five capabilities equips organizations to derive value from ECC applications, but then companies must apply those capabilities. We've found that four practices in particular help them do that, creating the conditions for a given application—and its underlying AI algorithm—to deliver on its promise.

Develop clear, realistic use cases. A use case provides a clear definition of what an ECC application will do and illustrates how its AI algorithms will enhance the execution and outcomes of a business process or set of processes. It shows how work will be divided between an application and a user.

In doing so, a use case establishes the need for process changes and provides initial insights into any new capabilities users will need (as well as any skills that will no longer be needed). A well-designed use case also facilitates the estimation of the costs and benefits of the ECC application.

Consider an ECC application in a call center: Its use case might include a simple version of an AI algorithm that matches customer queries to resolutions. It would show what the algorithm

With constantly changing business conditions, the data used to create an AI algorithm becomes a less accurate reflection of reality over time. So applications' learning must be managed throughout their life cycles.

would do and what automated resolutions the application could provide to customers. It would also show that some queries would be passed along to call center representatives. Required work and capability changes for call center employees could be inferred as well. All that information would allow a domain expert to roughly gauge the challenges of adoption and estimate intended benefits in terms of reduced response time, reduced labor, fewer follow-up calls, greater customer satisfaction, or a combination of outcomes.

Developing an ECC use case that is grounded in reality is a team activity. It is primarily the responsibility of domain experts and data scientists, who specify how an AI algorithm will enhance organizational outcomes and what data is needed to create it. But enterprise architects weigh in, too, identifying any new structures, roles, and systems required by a proposed ECC application, especially those affected indirectly by the new application. IT experts assess the need for integration with other applications and identify any additional IT support the application might require.

Properly developed use cases can help companies avoid sloppy or ill-considered ECC implementations that waste resources and may limit enthusiasm for—and effective implementation of—ECC. In fact, if the use cases for early ECC applications highlight quick wins for high-profile issues, they can be a powerful driver of organizational uptake of ECC. Bench scientists at one pharmaceutical company suggested developing an ECC application that could mine patent data for a specific disease knowing that if it was successful, the application itself would serve as a use case for similar applications for other diseases—and it did. Sometimes the algorithms themselves can be substantially reused. Wipro developed a use case for new-customer verification in the financial services sector, in which the AI algorithm automated the extraction and interpretation of information from customers' financial documents. This use case gave rise to an ECC application in the engineering sector that extracted and interpreted information from digitized blueprints.

Manage ECC application learning. AI algorithms in products such as smartphones use the data they process to improve themselves without human intervention. In contrast, ECC applications have a much more complex feedback loop. Business conditions and demands change constantly. As a result, the data used to create an AI algorithm becomes a less accurate reflection of reality over time—the algorithm drifts. It thus becomes necessary to manage the learning of the ECC applications throughout their life cycles.

Algorithm drift may occur quickly, as in predicting the sales of fashion apparel, or slowly, as in predicting the presence of a disease. To manage drift and keep ECC applications up-to-date, companies usually rely on a combination of IT backbone capabilities, data science competence, and domain proficiency. They build reporting mechanisms into ECC applications that generate alerts if the business results derived from the application's outputs are no longer aligned with the organization's goals, the algorithm's recommendations aren't within preestablished error ranges, or the application isn't running properly.

When deviations occur, AI algorithms need to be retrained and ECC applications relaunched. Domain experts and data scientists need to work together to identify, access, clean, tag, and architect new sources of data to improve the accuracy of AI algorithms and the utility of ECC applications. In addition, as the performance of the algorithm is better understood or as users become more proficient with the application, new business rules or processes that can enhance the value of the application may be required.

At OneBankAssure, domain experts and data scientists identified new external sources of data that could help identify productive sales leads, so they retrained their AI algorithm. They also learned that agent experience affected sales success, so they developed more elaborate rules to govern how the ECC application presented leads. The new data and business rules led to a richer, more complex ECC application that OneBankAssure continues to enhance.

Cocreate throughout the application life cycle. A data scientist or business domain expert cannot develop and sustain an ECC application in isolation. Interviewees in companies that effectively exploited AI repeatedly told us that they had, at first, badly underestimated the intense level of interdisciplinary cocreation needed to achieve success with ECC. They said they began to make progress only when they realized that ECC applications require people from disparate specialties and disciplines to work as a single team, not just during initial development and implementation but also in ongoing development throughout the application life cycle.

One reason cocreation is important for ECC is because business experts do not yet understand what AI can and can't do. During the development of ECC, close and sustained collaborative relationships across diverse areas of expertise can ameliorate this problem. At OneBankAssure, as a matter of hard-learned policy, every ECC application is created by a team of process owners and users with domain expertise, enterprise architects, and data scientists, with added assistance from the IT function.

There are few handoffs within the team. No team member ever works completely alone, and in the end, no one team member is responsible for success or failure. The interaction of team members results in a shared vocabulary about the business need and potential solutions, enabling them to better visualize and make sense of how people will actually use the application.

During implementation, owners of the IT operational backbone get involved not with a single handoff but rather by working with the ECC application team to cocreate a solution for integrating the application with the backbone upon production. After implementation, responsibility for maintaining and sustaining ECC applications continues to be highly interdependent in nature, as described above.

Think "cognitive". Companies that successfully develop and use ECC applications champion the uptake of AI and create positive buzz and excitement around its use. They encourage employees to generate ideas for new ECC applications that can improve their own work.

The employee response to ECC varies widely. Some people do not see the potential of ECC at first. Others have exaggerated expectations, thinking that ECC applications will automatically solve difficult business problems. Still others do not trust AI and see risks to ECC-enabled business processes, such as rogue behavior in AI algorithms and capability or job losses.

Domain experts who have seen what AI can do are the best stewards of realistic and credible conversations about ECC within their companies. Because of their business focus, they are more likely to be able to create a positive buzz around ECC than data scientists and IT professionals, who may be perceived as overly enamored with AI. Indeed, at Wipro, domain experts

Because of their business focus, domain experts are often more persuasive ECC champions than are data scientists and IT professionals, who may be perceived as overly enamored with AI.

were enlisted as AI champions—conducting "walkabouts" in their various departments, evangelizing ECC, and listening to ideas put forth by their colleagues.

The most likely sources of ideas for new ECC applications are people with domain proficiency or data science competence (or both). At OneBankAssure, operational managers spent several months in discussions with data science professionals to envision how their business might be affected by AI in the future, to develop ideas for new ECC applications, and to draft road maps for how their ideas could be developed and commercialized.

Proactive data science leaders also can be effective idea generators. At a pharmaceutical company we studied, one ECC project got its start at a lunch in which a business leader told a data scientist about a business problem, and the data scientist proposed a simple solution leveraging an already developed AI algorithm. In another company, the head of the data science unit organized seminars for functional and business leaders to identify areas in which ECC applications could best serve them.

The digital inquisitiveness of the entire workforce should be harnessed, too. Wipro, for example, crowdsources ideas from employees. It encourages them to envision and suggest new ECC applications, evaluating the ideas for their potential contribution to top-line growth, bottom-line profits, customer satisfaction, or employee satisfaction.

Business applications of AI may not create the same buzz as a self-driving car, but they can generate handsome returns—dramatic improvements in performance, profitability, revenues, and customer satisfaction. By cultivating the five capabilities and applying the four practices described in this chapter, business leaders can splice the ECC gene into their organizational DNA and set themselves up to reap those rewards.

It's a virtuous cycle: The capabilities enable employees to execute the practices, and the practices themselves exercise and strengthen the capabilities. This cycle helps companies become ever more adept at developing and using ECC applications that improve operations and create business value.

Learning through Doing

As companies apply their enterprise cognitive computing capabilities through the four key practices, they're also enriching their capabilities. Practices are, after all, opportunities to practice.

The pharmaceutical company we studied offers a good example. Recognizing that data science and ECC applications would become increasingly important to curing and preventing disease, the company hired data scientists to conduct workshops that would help senior staff (mainly business domain experts and enterprise architects) imagine the possibilities. They worked with business leaders to identify information-processing bottlenecks that created backlogs in drug discovery, clinical trials, manufacturing, and commercialization. The bottlenecks highlight opportunities for AI applications that could solve problems for small groups of analysts and decision makers in the organization.

These early efforts generated incremental business value, but the business leaders were far more focused on building capabilities than on building game-changing applications. They carefully chose use cases to meet the needs of people who naturally think "cognitive" and then engaged all the needed expertise—data scientists, domain experts, and IT specialists—to cocreate and manage the applications. In those pockets of the company, people deepened their understanding of organizational impacts and developed the capabilities to identify and pursue more ambitious ECC applications. What's more, the gains they made in efficiency and productivity inspired others in the company to seek out their own use cases and build their own capabilities. Creating this virtuous cycle of continuous organizational learning has mitigated the risks of the company's AI investments and positioned the company to make ECC a competitive advantage.

About the Research

The research activities on which this article is based were undertaken between January 2016 and December 2017, and covered companies across industries in North America, Europe, Asia, and Australia. We interviewed senior executives in IT and innovation units in 33 companies, as well as industry and technical experts in eight enterprise cognitive computing developer/vendor organizations, regarding ECC uptake in a range of organizations and industries, and ECC challenges and opportunities. We studied 51 ECC use cases (37% deployed; 48% in ideation or design stages; and 15% abandoned prior to development). We surveyed senior IT and technology leaders in 106 companies about ECC applications in place, application development and management issues, and outcomes. Finally, we researched and prepared three in-depth corporate case studies, for which we interviewed 35 people: C-level officers; functional leaders in IT, marketing, sales, and strategy; and data science and domain/process experts.[i]

---

i C. M. Beath, M. Tarafdar, and J. W. Ross, "OneBankAssure: Customer Intimacy through Machine Learning," working paper, MIT Center for Information Systems Research, Cambridge, Massachusetts, March 12, 2018; M. Tarafdar and C. M. Beath, "Wipro Limited: Developing a Cognitive DNA," working paper, MIT Center for Information Systems Research, Cambridge, Massachusetts, April 27, 2018; and J. W. Ross, K. Moloney, and C. M. Beath, "Pharmco: Becoming a Data-Science Driven Company," working paper, MIT Center for Information Systems Research, Cambridge, Massachusetts, February 21, 2019.

III

Trust and Ethics

12

Every Leader's Guide to the Ethics of AI

Thomas H. Davenport and Vivek Katyal

As artificial intelligence–enabled products and services enter our everyday consumer and business lives, there's a big gap between how AI can be used and how it should be used. Until the regulatory environment catches up with technology (if it ever does), leaders of all companies are on the hook for making ethical decisions about their use of AI applications and products.

Ethical issues with AI can have a broad impact. They can affect the company's brand and reputation, as well as the lives of employees, customers, and other stakeholders. One might argue that it's still early to address AI ethical issues, but our surveys and others suggest that about 30% of large companies in the US have undertaken multiple AI projects, with smaller percentages outside the US, and there are now more than 2,000 AI startups. These companies are already building and deploying AI applications that could have ethical effects.

Many executives are beginning to realize the ethical dimension of AI. A 2018 survey by Deloitte of 1,400 US executives knowledgeable about AI found that 32% ranked ethical issues as one of the top three risks of AI. However, most organizations don't yet have specific approaches to deal with AI ethics. We've identified seven actions that leaders of AI-oriented companies—regardless of their industry—should consider taking as they walk the fine line between can and should.

Make AI Ethics a Board-Level Issue

Since an AI ethical mishap can have a significant impact on a company's reputation and value, we contend that AI ethics is a board-level issue. For example, Equivant (formerly Northpointe), a company that produces software and machine learning-based solutions for courts, faced considerable public debate and criticism about whether its COMPAS systems for parole recommendations involved racially oriented algorithmic bias. Ideally, consideration of such issues would fall under a board committee with a technology or data focus. Unfortunately, these are relatively rare, in which case the entire board should be engaged.

Some companies have governance and advisory groups made up of senior cross-functional leaders to establish and oversee governance of AI applications or AI-enabled products, including their design, integration, and use. Farmers Insurance, for example, established two such boards—one for IT-related issues and the other for business concerns. Along with the board, governance groups such as these should be engaged in AI ethics discussions, and perhaps lead them as well.

A key output of such discussions among senior management should be an ethical framework for how to deal with AI. Some companies that are aggressively deploying AI, like Google, have developed and published such a framework.

Promote Fairness by Avoiding Bias in AI Applications

Leaders should ask themselves whether the AI applications they use treat all groups equally. Unfortunately, some AI applications, including machine learning algorithms, put certain groups at a disadvantage. This issue, called algorithmic bias, has been identified in diverse contexts, including judicial sentencing, credit scoring, education curriculum design, and hiring decisions. Even when the creators of an algorithm have not intended any bias or discrimination, they and their companies have an obligation to try to identify and prevent such problems and to correct them upon discovery.

Ad targeting in digital marketing, for example, uses machine learning to make many rapid decisions about what ad is shown to which consumer. Most companies don't even know how the algorithms work, and the cost of an inappropriately targeted ad is typically only a few cents. However, some algorithms have been found to target high-paying job ads more to men, and others target ads for bail bondsmen to people with names more commonly held by African Americans. The ethical and reputational costs of biased ad-targeting algorithms, in such cases, can potentially be very high.

Of course, bias isn't a new problem. Companies using traditional decision-making processes have made these judgment errors, and algorithms created by humans are sometimes biased as well. But AI applications, which can create and apply models much faster than traditional analytics, are more likely to exacerbate the issue. The problem becomes even more complex when black-box AI approaches make interpreting or explaining the model's logic difficult or impossible. While full transparency of models can help, leaders who consider their algorithms a competitive asset will quite likely resist sharing them.

Most organizations should develop a set of risk-management guidelines to help management teams reduce algorithmic bias within their AI or machine learning applications. They should address such issues as transparency and interpretability of modeling approaches, bias in the underlying data sets used for AI design and training, algorithm review before deployment, and actions to take when potential bias is detected. While many of these activities will be performed by data scientists, they will need guidance from senior managers and leaders in the organization.

Lean Toward Disclosure of AI Use

Some tech firms have been criticized for not revealing AI use to customers—even in prerelease product demos as with Google's AI conversation tool Duplex, which now discloses that it is an automated service. Nontechnical companies can learn from their experience and take preventive steps to reassure customers and other external stakeholders.

A recommended ethical approach to AI usage is to disclose to customers or affected parties that it is being used and provide at least some information about how it works. Intelligent agents or chatbots should be identified as machines. Automated decision systems that affect customers—say, in terms of the price they are being charged or the promotions they are offered—should reveal that they are automated and list the key factors used in making decisions. Machine learning models, for example, can be accompanied by the key variables used to make a particular decision for a particular customer. Every customer should have the right to an explanation—not just those affected by the

General Data Protection Regulation (GDPR) in Europe, which already requires it.

Also consider disclosing the types and sources of data used by the AI application. Consumers who are concerned about data misuse may be reassured by full disclosure, particularly if they perceive that the value they gain exceeds the potential cost of sharing their data.

While regulations requiring disclosure of data use are not yet widespread outside of Europe, we expect that requirements will expand, most likely affecting all industries. Forward-thinking companies will get out ahead of regulation and begin to disclose AI usage in situations that involve customers or other external stakeholders.

Tread Lightly on Privacy

AI technologies are increasingly finding their way into marketing and security systems, potentially raising privacy concerns. Some governments, for example, are using AI-based video surveillance technology to identify facial images in crowds and social events. Some tech companies have been criticized by their employees and external observers for contributing to such capabilities.

As nontech companies potentially increase their use of AI to personalize ads, websites, and marketing offers, it's probably only a matter of time before these companies feel pushback from their customers and other stakeholders about privacy issues. As with other AI concerns, full disclosure of how data is being obtained and used could be the most effective antidote to privacy concerns. The pop-up messages saying "our website uses cookies," a result of the GDPR legislation, could be a useful model for other data-oriented disclosures.

Financial services and other industries increasingly use AI to identify data breaches and fraud attempts. Substantial numbers of false-positive results mean that some individuals—both customers and employees—may be unfairly accused of malfeasance. Companies employing these technologies should consider using human investigators to validate frauds or hacks before making accusations or turning suspects over to law enforcement. At least in the short run, AI used in this context may actually increase the need for human curators and investigators.

Help Alleviate Employee Anxiety

Over time, AI use will probably affect employee skill sets and jobs. In the 2018 Deloitte survey of AI-aware executives, 36% of respondents felt that job cuts from AI-driven automation rise to the level of an ethical risk. Some early concerns about massive unemployment from AI-driven automation have diminished, and now many observers believe that AI-driven unemployment is quite likely to be marginal over the next couple of decades. Given that AI supports particular tasks and not entire jobs, machines working alongside humans seems more probable than machines replacing humans. Nonetheless, many workers who fear job loss may be reluctant to embrace or explore AI.

An ethical approach is to advise employees of how AI may affect their jobs in the future, giving them time to acquire new skills or seek other employment. As some have suggested, the time for retraining is now. Bank of America, for example, determined that skills in helping customers with digital banking will probably be needed in the future, so it has developed a program to train some employees threatened by automation to help fill this need.

Recognize That AI Often Works Best With – Not Without – Humans

Humans working with machines are often more powerful than humans or machines working alone. In fact, many AI-related problems are the result of machines working without adequate human supervision or collaboration. Facebook, for example, has announced it will add 10,000 additional people to its content review, privacy, and security teams to augment AI capabilities in addressing challenges with "fake news," data privacy, biased ad targeting, and difficulties in recognizing inappropriate images.

Today's AI technologies cannot effectively perform some tasks without human intervention. Don't eliminate existing, typically human, approaches to solving customer or employee problems. Instead—as the Swedish bank SEB did with its intelligent agent, Aida—introduce new capabilities as beta or trainee offerings and encourage users to provide feedback on their experience. Over time, as AI capabilities improve, communications with users may become more confident.

See the Big Picture

Perhaps the most important AI ethical issue is to build AI systems that respect human dignity and autonomy and reflect societal values. Google's AI ethics framework, for example, begins with the statement that AI should "be socially beneficial." Given the uncertainties and fast-changing technologies, it may be difficult to anticipate all the ways in which AI might impinge on people and society before implementation—although certainly companies should try to do so. Small-scale experiments may uncover negative outcomes before they occur on a broad scale. But when signs of harm appear, it's important to acknowledge and act on emerging threats quickly.

Of course, many companies are still very early in their AI journeys, and relatively few have seriously addressed the ethics of AI use in their businesses. But as bias, privacy, and security issues become increasingly important to individuals, AI ethical risks will grow as an important business issue that deserves a board-level governance structure and process.

13

When People Don't Trust Algorithms

Berkeley J. Dietvorst, interviewed by Paul Michelman

Even when faced with evidence that an algorithm will deliver better results than human judgment, we consistently choose to follow our own minds.

Why?

MIT Sloan Management Review editor in chief Paul Michelman sat down with Berkeley J. Dietvorst, assistant professor of marketing at the University of Chicago Booth School of Business, to discuss a phenomenon Dietvorst has studied in great detail. What follows is an edited and condensed version of their conversation.

***MIT Sloan Management Review*: What prompted you to investigate people's acceptance or lack thereof of algorithms in decision-making?**

Berkeley Dietvorst: When I was a PhD student, some of my favorite papers were old works by [the late psychology scholar and behavioral decision research expert] Robyn Dawes showing that algorithms outperform human experts at making certain types of predictions. The algorithms that Dawes was using were very simple and oftentimes not even calibrated properly.

A lot of others followed up Dawes's work and showed that algorithms beat humans in many domains—in fact, in most of the domains that have been tested. There's all this empirical work showing algorithms are the best alternative, but people still aren't using them.

So we have this disconnect between what the evidence says people should do and what people are doing, and no one was researching why.

What's an example of these simple algorithms that were already proving to be superior?

Dietvorst: One of the areas was predicting student performance during an admission review. Dawes built a simple model: Take four or five variables—GPA, test scores, etc.—assign them equal weight, average them on a numerical scale, and use that result as your prediction of how students will rank against each other in actual performance. That model—which doesn't even try to determine the relative value of the different variables—significantly outperforms admissions experts in predicting a student's performance.

What were the experiments you conducted to try to get at the reasons we resist algorithms?

Dietvorst: We ran three sets of experiments.

For the first paper, we ran experiments where the participants' job was to complete a forecasting task, and they were incentivized to perform well. The better they performed, the more money they would earn in each experiment. There were two stages: first a practice round—for both humans and algorithms—and then a stage where participants were paid based on the quality of their performance.

In the practice round, we manipulated what forecasts participants were exposed to. Some made their own forecasts and saw those of the algorithm. Some made only their own forecasts. Some saw only the algorithm's results. Some saw neither. So each group had different information about how well each forecasting option had performed during the practice round.

For the second stage, participants could choose to forecast the results themselves or rely on the algorithm. The majority of participants who had not seen the algorithm's results from the first round chose to use it in the second round. However, those people who had seen the algorithm's results were significantly less likely to use it, even if it beat their own performance.

Once people had seen the algorithm perform and learned that it was imperfect, that it makes mistakes, they didn't want to use it. But there wasn't a similar effect for them. Once I made a forecast and learned that I was imperfect, I wasn't less likely to use my own forecast. We saw that effect only for the algorithm.

And for the second experiment?

Dietvorst: In the second paper, we tried to address the problem: How can we get people to use algorithms once they know that they're imperfect?

We began with the same basic question for participants: human or algorithm? In these experiments, however, there was an additional twist. Some participants were given the choice between using the algorithm as it existed or not at all. Other participants, if they chose to use the algorithm, could make some adjustments to it.

We found that people were substantially more willing to use algorithms when they could tweak them, even if just a tiny amount. People may be unwilling to use imperfect algorithms as they exist—even when the algorithm's performance has been demonstrated superior to their own—but if you give the person any freedom to apply their own judgment through small adjustments, they're much more willing.

So those are the key findings from the first two papers I wrote with my coauthors Joe Simmons and Cade Massey. Following on those, I have a solo paper where I'm investigating more about why people weren't willing to use algorithms once they learned that they're imperfect.

Most people in my experiment used human forecast by default, which positions the algorithm as an alternative. And the way they make the decision about whether or not to use the algorithm is by asking, "Will this algorithm meet my performance goal?" even if that goal is unrealistic for human forecasts, too. They don't choose the algorithm if it won't meet some lofty goal.

What they should more reasonably ask is, "Is this algorithm better than me?"—which it usually is. So people fail to ask the right question and end up holding the two options to different standards.

And to what do you attribute that?

Dietvorst: That's an interesting question. I'm not sure how this decision process came about or why people are making the decision this way. And I've found it's not actually unique to algorithms.

When choosing between two human forecasters, people do the same thing. If you assign them to have one forecaster as their default and you ask them how well would the other forecaster have to perform in order for you to switch, people say the other forecaster would have to meet my performance goals, just as with the algorithm.

It seems like people are naturally making what I would call the wrong comparison.

So it's kind of a switching cost?

Dietvorst: Not necessarily. The way I would think about a switching cost would be: I'm used to using human judgment, so an algorithm has to perform X percent better or X points better than me, or a human, for me to switch to it, right?

But that's not really how it works. People are comparing the alternative to their performance goal, rather than comparing the two options. So, the higher the performance goal I give you, the better you need the algorithm to perform in order to switch to it, even though your own performance is staying constant.

So it doesn't seem like a switching cost, at least as we tend to think of the term.

What I find so interesting is that it's not limited to comparing human and algorithmic judgment; it's my current method versus a new method, irrelevant of whether that new method is human or technology.

Dietvorst: Yes, absolutely. That's exactly what I've been finding.

I think one of the questions that's going to come up is, "Well, what do I do about this? Is simple recognition of the bias enough to counter it?"

Dietvorst: If I can convince someone that the right question to ask is, "Does this algorithm outperform what you're currently using?" instead of, "Does this algorithm meet some lofty performance goal?" and that person buys in and says, "Yes, you're right, I should use algorithms that outperform what I'm currently doing," then, yes, that would work. I don't know how easy or hard it would be to get people to buy into that, though.

And in a larger organization, thousands of decisions are being made every day. Without this bias being known, there really isn't an obvious corrective measure, is there?

Dietvorst: The studies I've done suggest a couple restrictions that could reduce the bias.

People are deciding whether or not to use the algorithm by comparing it to the performance goal that they have. If you incentivize people to attempt to deliver performance much better than an algorithm has shown it's capable of, it's not so surprising that they ditch the algorithm to chase down that incentive with human judgment—even if it's unrealistic they will achieve it.

If you lower their performance goal, the algorithm will be compared more favorably and people may be more likely to use it.

So the problem exists in situations where the goal itself is unreasonable.

Dietvorst: Yes, if you have some forecasting goal that is very hard to achieve and an algorithm hasn't achieved it in the past, then you could see how it would make sense, in a certain way, for people not to use the algorithm. They're pretty sure it's not going to achieve the goal. So they use human judgment and end up performing even worse than the algorithm.

Presumably, we're in an age now where the quality of algorithms is increasing – perhaps dramatically. I'm wondering whether this phenomenon will make our biases more or less pronounced. On the one hand, you could see the quality of algorithms catching up to people's reference points. But the inverse of that is the

reference point will continue to move at a speed as high if not higher than the ability of the algorithm.

Dietvorst: I agree: That could go either way. But I would like to push back a little bit on this idea that algorithms are really great. The literature shows that on average, when predicting human behavior, algorithms are about 10% to 15% better than humans. But humans are very bad at it. Algorithms are significantly better but nowhere near perfection. In many domains, I don't see any way that they're going to get close to perfection very soon.

There is a lot of uncertainty in the world that can't be resolved or reduced—that is unknowable. Like when you roll a die, you don't know what number is going to come up until it happens. A lot of that type of aleatory uncertainty is determining outcomes in the real world. Algorithms can't explain that.

Suppose Google Maps is telling you the fastest route to a new place. It can't predict if there's going to be a giant accident right in front of you when you're halfway there. And so, as long as there's random error and there's aleatory uncertainty that factors into a lot of these outcomes—which it does to a larger extent than people recognize—algorithms aren't going to be perfect, and they aren't really even going to be close to perfect. They'll just be better than humans.

So what's next? Is this an ongoing field of study for you?

Dietvorst: Absolutely. There's a lot more to understand about how people think algorithms operate; what they think are the differences between algorithms and humans; and how that affects their use of algorithms. There's still really interesting research to be done.

14

The Risk of Machine Learning Bias (And How to Prevent It)

Chris DeBrusk

Many companies are turning to machine learning to review vast amounts of data, from evaluating credit for loan applications, to scanning legal contracts for errors, to looking through employee communications with customers to identify bad conduct. New tools allow developers to build and deploy machine learning engines more easily than ever: Amazon Web Services recently launched a "machine learning in a box" offering called SageMaker, which non-engineers can leverage to build sophisticated machine learning models, and Microsoft Azure's machine learning platform, Machine Learning Studio, doesn't require coding.

But while machine learning algorithms enable companies to realize new efficiencies, they are as susceptible as any system to the "garbage in, garbage out" syndrome. In the case of self-learning systems, the type of "garbage" is biased data. Left unchecked, feeding biased data to self-learning systems can lead to unintended and sometimes dangerous outcomes.

In 2016, for example, an attempt by Microsoft to converse with millennials using a chat bot plugged into Twitter famously created a racist machine that switched from tweeting that "humans are super cool" to praising Hitler and spewing out misogynistic remarks. This scary conclusion to a one-day experiment resulted from a very straightforward rule about machine learning—the models learn exactly what they are taught. Correctional Offender Management Profiling for Alternative Sanctions (COMPAS), a machine learning system that makes recommendations for criminal sentencing, is also proving imperfect at predicting which people are likely to reoffend because it was trained on incomplete data. Its training model includes race as an input parameter, but not more extensive data points like past arrests. As a result, it has an inherent racial bias that is difficult to accept as either valid or just.

These are just two of many cases of machine learning bias. Yet there are many more potential ways in which machines can be taught to do something immoral, unethical, or just plain wrong.

Best Practices Can Help Prevent Machine Learning Bias

These examples serve to underscore why it is so important for managers to guard against the potential reputational and regulatory risks that can result from biased data, in addition to figuring out how and where machine learning models should be deployed to begin with. Best practices are emerging that can help to prevent machine learning bias. Below, we examine a few.

Consider bias when selecting training data. Machine learning models are, at their core, predictive engines. Large data sets train machine learning models to predict the future based on the past. Models can read masses of text and understand intent, where intent is known. They can learn to spot differences—between, for instance, a cat and a dog—by consuming millions of pieces of data, such as correctly labeled animal photos.

The advantage of machine learning models over traditional statistical models is their ability to quickly consume enormous numbers of records and thereby more accurately make predictions. But since machine learning models predict exactly what they have been trained to predict, their forecasts are only as good as the data used for their training.

For example, a machine learning model designed to predict the risk of business loan defaults may advise against extending credit to companies with strong cash flows and solid management teams if it draws a faulty connection—based on data from loan officers' past decisions—about loan defaults by businesses run by people of a certain race or in a particular ZIP code. A machine learning model used to scan reams of résumés or applications to schools might mistakenly screen out female applicants if the historical data used to train it reflects past decisions that resulted in few women being hired or admitted to a college.

These types of biases are especially pervasive in data sets based on decisions made by a relatively small number of people. As a best practice, managers must always keep in mind that if humans are involved in decisions, bias always exists—and the smaller the group, the greater the chance that the bias is not overridden by others.

Root out bias. To address potential machine learning bias, the first step is to honestly and openly question what preconceptions could currently exist in an organization's processes, and actively hunt for how those biases might manifest themselves in data. Since this can be a delicate issue, many organizations bring in outside experts to challenge their past and current practices.

Once potential biases are identified, companies can block them by eliminating problematic data or removing specific components of the input data set. Managers for a credit card company, for example, when considering how to address late payments or defaults, might initially build a model with data such as ZIP codes, type of car driven, or certain first names—without acknowledging that these data points can correlate with race or gender. But that data should be stripped, keeping only data directly relevant to whether or not customers will pay their bills, such as data on credit scores or employment and salary information. That way, companies can build a solid machine learning model to predict likelihood of payment and determine which credit card customers should be offered more flexible payment plans and which should be referred to collection agencies.

A company can also expand the training data set with more information to counterweight potentially problematic data. Some companies, for example, have started to include social media data when evaluating the risk of a customer or client committing a financial crime. A machine learning algorithm may flag a customer as high risk if he or she starts to post photos on social media from countries with potential terrorist or money-laundering connections. This conclusion can be tested and overridden, though, if a user's nationality, profession, or travel proclivities are included to allow for a native visiting his or her home country or a journalist or businessperson on a work trip.

Regardless of which approach is used, as a best practice, managers must not take data sets at face value. It is safe to assume that bias exists in all data. The question is how to identify it and remove it from the model.

Counter bias in "dynamic" data sets. Another challenge for machine learning models is to avoid bias where the data set is dynamic. Since machine learning models are trained on events that have already happened, they cannot predict outcomes based on behavior that has not been statistically measured. For example, even though machine learning is extensively used in fraud detection, fraudsters can outmaneuver models by devising new ways to steal or escape detection. Employees can hide bad behavior from machine learning tools used to identify bad conduct by using underhanded techniques like conversing in code.

To attempt to draw new conclusions from current information, some companies use more experimental, cognitive, or artificial intelligence techniques that model potential scenarios. For example, to outsmart money launderers, banks may conduct so-called war games with ex-prosecutors and investigators to discover how they would beat their system. That data is then used to handcraft a more up-to-date machine learning algorithm.

But even in this situation, managers risk infusing bias into a model when they introduce new parameters. For example, social media data, such as pictures posted on Facebook and Twitter, is increasingly being used to drive predictive models. But a model that ingests this type of data might introduce irrelevant biases into its predictions, such as correlating people wearing blue shirts with improved creditworthiness.

To avoid doing so, managers must ensure that the new parameters are comprehensive and empirically tested—another best practice. Otherwise, those parameters might skew the model, especially in areas where data is poor. Insufficient data could impact, say, credit decisions for classes of borrowers who a bank has never lent to previously but wants to in the future.

Balance transparency against performance. One temptation with machine learning is to throw increasingly large amounts of data at a sophisticated training infrastructure and allow the machine to figure it out. For example, public cloud companies have recently released comprehensive tools that use automated algorithms instead of an expert data scientist to train and determine the parameters intended to optimize machine learning models.

While this is a powerful method for building complex predictive algorithms quickly and at lower cost, it also comes with the downside of limited visibility and the risk of the "machine running wild" and having an unconscious bias due to training data that is extraneous (like the blue shirt bias described above). The other challenge is that it is very difficult to explain how complex machine learning models actually work, which is problematic in industries that are heavily regulated.

One of the potential options to address this risk is to take a staged approach to increasing the sophistication of the model and making a conscious decision to progress at every stage.

A good example is a process used by a major bank in building a model that attempted to predict whether mortgage customers were about to refinance, with the goal of making a direct offer to them and ideally retaining their business. The bank started with a simple regression-based model that tested its ability to predict when customers would refinance. It then created a set of more sophisticated "challenger" models that used more advanced machine learning techniques and were more precise. By confirming that the challenger models were more accurate than the base regression model, bank managers became comfortable that their more complex and opaque machine learning approach was operating in line with expectations and not propagating unintended biases. The process also enabled them to verify that the machine learning tool's balance between transparency and sophistication was in line with what is expected in the highly regulated financial services industry.

Careful Planning Is a Necessity

It is tempting to assume that, once trained, a machine learning model will continue to perform without oversight. In reality, the environment in which the model is operating is constantly changing, and managers need to periodically retrain models using new data sets.

Machine learning is one of the most exciting technical capabilities with real-world business value in the last decade. When combined with big-data technology and the massive computing capability available via the public cloud, machine learning promises to change how people interact with technology, and potentially entire industries. But as promising as machine learning technology is, it requires careful planning to avoid unintended biases.

Creators of the machine learning models that will drive the future must consider how bias might negatively impact the effectiveness of the decisions the machines make. Otherwise, managers risk undercutting machine learning's potentially positive benefits by building models with a biased mind of their own.

15

Even If AI Can Cure Loneliness, Should It?

David Kiron and Gregory Unruh

Many experts believe augmentation and automation are the shining stars of business uses of AI; their promise of greater productivity has lit up the executive imagination.

In their shadow, however, a growing number of AI applications and devices are helping humans satisfy a basic need to connect with others. In particular, markets are slowly forming around artificially intelligent, emotionally attuned, responsive robots that people can relate to as companions. Many people can form emotional connections with their bots, either in lieu of human alternatives or in addition to them.

Where there is an unfilled human need, there is a business opportunity. Large-scale social problems, like the global loneliness epidemic, are driving demand for robot companions. The AARP estimates that one out of three US adults older than age 45 suffer from chronic loneliness. In Britain, researchers estimate that 9 million adults are often or always lonely; one out of three adults over 75 say their feelings of loneliness are out of control. In January 2018, Britain named a loneliness minister after recognizing its serious, multibillion-dollar toll on the UK economy. Loneliness is associated with premature death, productivity loss, and various health costs. More than a dozen startups are developing robot home companions. While some are struggling, there is little question that demand for these products exists, and is likely to grow stronger with an aging population.

Eldercare "is rapidly becoming one of the most daunting health care challenges of our day," says former Harvard Medical School professor William A. Haseltine. An NIH-funded Census Bureau report estimates that by 2050, nearly 17% of the world's population, or 1.6 billion people, will be at least age 65, double the percentage of today. Many companies are developing robots to provide services to that growing cohort, such as making schedule recommendations, offering medicine reminders, and coordinating care. Although most of these aren't designed specifically with loneliness in mind, they do provide companionship, which many of the elderly desperately need.

So technology isn't just a cause for the loneliness epidemic, as many suggest. It's also possibly a solution. The early popularity of social robots suggests that there's quite a bit of pent-up demand for nonpharmaceutical alternatives—not only to address loneliness once it sets in but also to stave it off in the first place.

Of course, the human need for connection is physical as well as emotional. Market demand for social robots that satisfy sexual appetites is also on the rise. You thought Tinder was bad: Sex with your own robot is now an option, and thousands of people are taking advantage. Entrepreneurs are combining advances in materials science, robotics, sensor technology, and natural language processing to create anatomical simulacra that provide physical pleasure. Granted, this approach to satisfying a basic need may not be long lasting for a given person. Some may doubt whether it constitutes relating at all. Even so, with no regulatory structures around sexbots—and none in the making—the effects of this growth industry will inevitably rejigger social norms. Noel Sharkey, from the Foundation for Responsible Robotics, says: "We're just doing all this stuff with machines because we can, and not really thinking how this could change humanity completely." The makers of some robotic sex dolls are looking to expand sales to eldercare facilities; sex doll brothels already exist in some countries.

The market opportunities for social AI extend beyond caring for the lonely and the elderly: Collaborative robots (aka "cobots") provide a substitute for traditional social connection in the workplace (for example, SoftBank's Pepper is used as a customer service aid in hotels). AI-driven online games like Fortnite or dating sites like Match.com rely on algorithms to help like-minded people find and connect with one another so that they don't have to do the work themselves, which can be excruciating for the introverted.

Understanding social AI as a market maker is critical for company strategists, as well as product developers. But it's equally important to recognize that, for better and worse, the social arrangements we take for granted today are also at stake: The roles people play in their own and others' lives are increasingly mediated by technological third-party actors with greater emotional, linguistic, and social sophistication. The implications are both daunting and clear: If part of what makes us human is to connect emotionally with others, and technology increasingly plays the role of emotional connector, what it means to be distinctively "human" becomes a much more complicated question.

The societal ramifications of social AI products are not entirely for regulators to address. Makers of social AI products and services need to tread carefully and really consider the difference between "Can we make and sell this?" and "Should we make and sell this?" The question is "Will executives looking to address real and pressing social ills with social AI work effectively with regulators to address unintended consequences of their business efforts?"

16

AI Can Help Us Live More Deliberately

Julian Friedland

As I search online for a present for my mother, considering the throw pillows with sewn-in sayings, plush bathrobes, and other options, and eventually narrowing in on one choice over the others, who exactly has done the deciding? Me? Or the algorithm designed to provide me with the most "thoughtful" options based on a wealth of data I could never process myself? And if Mom ends up hating the embroidered floral weekender bag I end up "choosing," is it my fault? It's becoming increasingly difficult to tell, because letting AI think for us saves us the trouble of doing it ourselves and owning the consequences.

AI is an immensely powerful tool that can help us live and work better by summoning vast amounts of information. It spares us from having to undergo many mundane, time-consuming, nerve-wracking annoyances. The problem is that such annoyances also play a key adaptive function: They help us learn to adjust our conduct in relation to one another and the world around us. Engaging directly with a grocery bagger, for instance, forces us to confront his or her humanity, and the interaction (ideally) reminds us not to get testy just because the line isn't moving as quickly as we'd like. Through the give and take of such encounters, we learn to temper our impulses by exercising compassion and self-control. Our interactions serve as a constantly evolving moral-checking mechanism.

Similarly, our interactions within the wider world of physical objects forces us to adapt to new environments. Walking, bicycling, or driving in a crowded city teaches us how to compensate for unforeseen obstacles such as varying road and weather conditions. On countless occasions every day, each of us seeks out an optimal compromise between shaping ourselves to fit the world and shaping the world to fit ourselves.[1] This kind of adaptation has led us to become self-reflective, capable of ethical considerations and aspirations.

Our rapidly increasing reliance on AI takes such interactions out of our days. The frictionless communication AI tends to propagate may increase cognitive and emotional distance, thereby letting our adaptive resilience slacken and our ethical virtues atrophy from disuse.[2] Relying on AI to pre-select gifts for friends and family, for example, spares us the emotional labor of considering their needs and wants in our ordinary interactions with them to select a genuinely thoughtful gift. Many trends already well underway involve the offloading of cognitive, emotional, and ethical labor to AI software in myriad social, civil, personal, and professional contexts.[3] Gradually, we may lose the inclination and capacity to engage in critically reflective thought, making us more cognitively and emotionally vulnerable and thus more anxious[4] and prone to manipulation from false news, deceptive advertising, and political rhetoric.

In this article, I consider the overarching features of this problem and provide a framework to help AI designers tackle it through system enhancements in smartphones and other products and services in the burgeoning internet of things (IoT) marketplace. The framework is informed by two ideas: psychologist Daniel Kahneman's cognitive dual process theory[5] and moral self-awareness theory, a four-level model of moral identity that I developed with Benjamin M. Cole, a professor at Fordham University's Gabelli School of Business.[6]

Theories of Mind in an AI World

Cognitive dual process theory describes two overarching decision-making processes: (1) the autonomous mind, which automatically reacts to stimuli, and (2) the reflective mind, which responds consciously in a deliberate and reasoned fashion.[i]

Most AI-assisted platforms function to free up the attention of the conscious reflective mind for any activities that immediately suit a person's interests or grab his or her attention. Ideally, each new outsourced task is accomplished more effectively than via direct unassisted interaction. Thus, AI allows us to conveniently increase the levels at which we may productively process incoming information from the external physical and social worlds.

AI systems typically guide users with visceral notices, which researchers have divided into three general categories:[ii]

- **Familiarity notices** use familiarity with one technology to inform users about another. Example: camera-clicking sounds and dial tones on smartphones.
- **Psychological reaction notices** use common psychological reactions to shape a consumer's conception of the product or service. Example: casual interface designs like friendly avatars that signal greater honesty and openness.
- **Showing notices** promote self-awareness by showing users the results of their activities. Example: screen-time data embedded in the iPhone iOS 12.

i D. Kahneman, *Thinking, Fast and Slow* (New York: Farrar, Straus and Giroux, 2011).

ii M. R. Calo, "Against Notice Skepticism in Privacy (and Elsewhere)," *Notre Dame Law Review* 87, no. 3 (October 2013): 1027–1072.

Familiarity notices and psychological reaction notices are designed to trigger only the autonomous mind, but showing notices introduce communicative friction designed to trigger the reflective mind. Screen-time software embedded in the iPhone operating system shows people how often they use social networking, entertainment, and productivity apps. This allows them to better understand and take control of their own behavior.

When Convenience Leads to Disengagement

The most immediately attractive feature of AI technology is its promise to handle the mundane aspects of life, thereby increasing the amount of time and attention each of us can devote to activities we consider more rewarding. Of course, every time this kind of outsourcing occurs, we cede a degree of control. Getting comfortable with these trade-offs reinforces new habitual behaviors that entail a measure of disengagement: from one another, the physical world, and even ourselves. This is because every time we delegate a degree of control to the AI system, we also invest a degree of trust into that system. In so doing, we will often shift from relying on what Kahneman calls our reflective mind (and its deliberative decision-making) to our autonomous mind (and its automatic reactions that guide decisions). This makes it easy to complete a routine task. But repeating this process creates a risk that our actions become increasingly automatic and less reflective overall, leading to six forms of disengagement:[7]

1. **Increased passivity** As we accept assistance to complete a task, we require less effort to carry it out. We may become spectators rather than active participants. The AI systems that Netflix, Amazon Prime, and Facebook use to preselect

entertainment and news options are examples. When we let these systems determine our options, we rarely confront perspectives that might challenge our preconceptions and biases. Gradually, we may become less prepared to expend the effort needed to think deeply and critically, thereby disengaging long-term memory.[8]

2. **Emotional detachment** Diminished participation leads to emotional disengagement. Consequently, our actions can become insincere or deceptive. Think of a customer call center, where an AI system in a help desk or sales context aggressively coaches agents in real time as they respond to customers' emotional cues.[9] Such software, ideally designed to train operators to become more sensitive to customers' concerns, could have the reverse effect, making us increasingly inured to emotional cues because we will have less practice picking up these cues ourselves and have less interest in doing so.
3. **Decreased agency** Disengagement reduces our power to make our own decisions by lessening our awareness of actions we might take. Consider an automated vehicle preprogrammed to weigh competing ethical priorities during a crash, such as whether to hit a pedestrian or another vehicle. Auto insurance rates might be adjusted according to the degree to which we set the automated driving system to integrate others' interests into the calculus.[10] And we would relinquish the agency to make our own choice as the crash takes place.
4. **Decreased responsibility** In ceding control over a decision-making process, we can become less accountable for results—whether they are good or bad—because responsibility is

diffused across the entire AI-based system, from design to delivery. Imagine a dieting app that orders prepared foods to be delivered to you according to a weight-loss plan set up by AI. If you lose weight, who deserves the credit? And if you don't, whose fault is it?[11]

5. **Increased ignorance** AI translates our wants into algorithmic shorthand or mechanical processes that may end up functioning differently than we would ourselves. Of course, that can make up for deficiencies in our knowledge—but it can also reinforce those deficiencies. Virtual navigational apps like those offered by Waze, Garmin, and others do not require you to acknowledge your surroundings. You might, for instance, keep circling an incorrect location that the mapping app has not yet updated, out of preferential bias for the AI system, instead of returning back to your own direct perceptions and judgments.[12] At your intended destination, you might have no idea what route you took to get there nor how to get back to where you started without AI assistance.
6. **De-skilling** Depending on an intermediary for completing routine tasks can dull many of the trained skills we rely on to interact with the physical world around us. We may forget how to perform basic tasks or become less proficient at doing them unaided. Using only navigation apps lulls us into forgetting how to use a conventional map or, in a future era of autonomous vehicles, even how to drive without the apps. We may also lose motivation to acquire new skills, opting instead for ever more outsourcing solutions.

Together, these trends present an ethical challenge: Because they multiply the instances in which we go through life while operating on autopilot, they have the potential to loosen our social bonds, exacerbate conflicts, and hamper moral progress by stifling self-critical thought. To mitigate these threats, designers of AI systems should build in features and interfaces that periodically re-trigger our reflective minds.

It Takes More Than "Nudges" to Make Us Think

In their influential book, *Nudge*, behavioral economist Richard Thaler and legal scholar Cass Sunstein have argued that cognitive nudges can spur us to action by using triggers that evoke emotions like empathy or self-interest.[13] Unfortunately, such nudges have limited power in practice because they prompt only behavioral impulses and do not engage critical reflection. This is the case even when pressing health risks are concerned. In a study of 1,509 patients who had heart attacks, efforts to prompt people to adhere to medication prescriptions (including electronic pill bottles and the chance for $5 or $50 rewards for enlisting the support of a friend or family member) did not significantly improve the likelihood that people would take their medicine.[14]

Triggering the reflective mind is more likely to solve the problem of disengagement and mitigate the risks of losing skills in the age of AI. By creating what we can call cognitive speed bumps that force us to reflect on decisions worthy of greater reflection, developers of AI systems can reintroduce interactive friction into the experiences they host. So as Mom's birthday approaches, instead of suggesting purchases, our AI system might instead suggest a good time to call or pay Mom a visit—an opportunity to enhance the personal relationship and even help come up with a thoughtful (and desired) gift.

The ramifications are profound. Perhaps the most seductive aspect of AI-assisted platforms is that they promote what technology ethicist Shannon Vallor describes as "frictionless interactions that deftly evade the boredom, awkwardness, conflict, fear, misunderstanding, exasperation, and uncomfortable intimacies that often arise from traditional communications, especially face-to-face encounters in physical space."[15] Here, Vallor is referring mainly to the avoidance of live conversations, through social media. But she may as well be talking about evasion of all the practical drudgeries of life, from reading a map, driving a car, and minding one's surroundings to making a grocery list, shopping, and cooking. And though most of us still have such frictional experiences, AI-assisted platforms promise to guide our attention in whatever directions we are likely to find most immediately satisfying, thereby reducing the chances that we will have to experience unpleasant friction. As a result, our moral attention—the ability to redirect our focus, delay gratification, temper our emotional urges, and restrain our unthinking reactions—erodes.

We need something to counteract this tendency: an AI choice architecture designed to preserve healthy measures of interactive friction between ourselves and the wider world.

How Friction Fosters Moral Self-Awareness

There is value to a world of friction-filled interactions. For instance, new research on childhood self-control suggests that one's cultural[16] and socio-economic[17] environments may play a far greater role than genetic factors in developing grit and perseverance, which are highly correlated with professional success later in life.[18] It is only by learning how to navigate interactions

It is only by learning how to navigate interactions that are not set up for our comfort that we are able to fully develop executive control over our own consciousness.

that are not set up for our comfort that we are able to fully develop executive control over our own consciousness.[19]

Such interactions also foster moral self-awareness. As we experience friction again and again, the ways we react to various stimuli change, and moral identity evolves: We begin to think and feel differently about what our actions say about ourselves.[20]

The social psychological literature has established a clear relationship between what's called the self-importance of moral identity and moral thought and action,[21] and the wider literature on civic-mindedness indicates that pride is the most effective moral motivator of civic behavior.[22] There is also evidence that ethical consumers are happier and have stronger repurchase intentions when motivated by their moral self-image than when motivated by emotions such as guilt and empathy.[23]

What does all this have to do with AI? Designers of AI systems can use the four levels of moral self-awareness described below as a guide for developing applications that encourage reflective behavior. By incorporating triggers for interactive friction, they can prompt users to consider how their actions reflect their personal values and help them ascend to higher levels of awareness.

Level 1: Social reflection. At this level, people rely chiefly upon negative feedback they receive from observers to guilt or shame them into changing their behavior. Researchers have demonstrated the power of negative feedback to inhibit a person's selfish behavior. For example, participants primed in a tragedy of the commons experiment to be self-interested gradually learned to temper their self-interest after being shamed by other subjects left with fewer resources.[24] Eventually, all subjects showed a preference for lowered individual returns in favor of equitable and sustainable longer-term outcomes.

It is only by learning how to navigate friction-filled interactions that we are able to fully develop executive control over own consciousness.

Level 2: Self-reflection. At this level, rather than relying on others' complaints to acknowledge the negative impacts of their actions, actors start to serve as their own source of feedback. This happens when they see the outcomes of others' behavior or when they consider the immediate ramifications of their own actions. For example, a person who notices a room containing swept litter is 2.5 times less likely to toss trash on the floor than in a litter-strewn room.[25] Observing the neatened-up litter increases the observer's propensity to keep the room clean.

Level 3: Anticipatory self-reflection. At this level, people start to anticipate potential negative consequences of their actions and do so independently from others' signals. This behavior often comes after self-reflection on prior behavior has led to an internal sense of guilt or shame. At a crucial turning point in the tragedy of the commons experiment mentioned above,[26] one participant asked aloud, "Are we bad people?" This question was not so much an effort to shame other group members as an attempt to reconcile the inconsistency between one's prior action (to serve self-interest) and one's aspirational moral self-image. Such a reflective moment represents a crucial step, one that reveals the moral obligations of individuals to shape themselves to fit the world and their own aspirations within it.

Level 4: Proactive self-reflection. At the highest level, people become increasingly forward-looking, considering both negative and positive impacts. They purposely engage in appropriate actions to realize positive outcomes. They internalize the self-image of potential hero rather than potential villain.[27] At best, these decisions are habit-forming, bringing people closer to becoming whom they aspire to be. This state of mind is linked with achieving greater happiness based on an individual's self-conception.[28]

Triggering the Reflective Mind

In traditional face-to-face interactions, the external physical or social world provides the friction necessary to trigger the reflective mind into modifying one's behavior for the better. As AI removes opportunities for those interactions, developers need a tool for tapping into users' moral self-awareness. Showing notices, a type of visceral notice that AI systems can incorporate to shape users' decision-making, can serve as that tool and compensate for the loss of give-and-take interactions in the social and physical world.

Showing notices provide users with snapshots of their behavior (the number of steps taken in a day, for example, or the amount of time spent online). They can enhance AI applications by encouraging users to move from the first, second, and third levels of moral self-awareness, in which negative feelings like guilt and shame primarily drive individual behaviors, toward Level 4, in which positive aspirations encourage people to act, conscious that their choices can make a difference for themselves and society. Enabling users to share their progress on a given issue with others in a social group further enhances an app's potential.

Considering that by current projections, global IoT spending could reach $1.4 trillion by 2021, such functionality presents rich opportunities for research and development.[29] Five lifestyle categories in particular have significant potential for this type of innovation: health and well-being, social responsibility, media and civic engagement, skill maintenance, and personal edification. We'll consider each one here.

By providing "showing notices" – snapshots of users' behavior – AI applications can encourage people to move toward the highest level of moral self-awareness, where positive aspirations drive individual behavior.

Health and well-being. There is already significant movement in providing showing notices in health and wellness apps—from those that facilitate personal fitness, mindfulness, or sleep management to those that allow us to set screen-time limits on our cell phones. Smart refrigerators are another frontier. For example, adding showing notices that illustrate patterns of consumption of highly processed, high-sugar, canned, frozen, and fresh food, along with daily calorie consumption data, could help users improve nutrition. Combined with data from grocery delivery services, such notices could guide users to order groceries according to healthier recipes and locally or sustainably sourced foods.

Social responsibility. Another area with potential is in helping people make thoughtful brand and investment choices that align with their social values. A few apps now highlight possible ethical concerns in financial portfolios, flagging sectors that users may wish to avoid in light of stated preferences (such as alcohol, petroleum, and tobacco) and providing finer-grained notices about any ethical quandaries companies may be involved in. Smart refrigerators could provide notices about the carbon footprint of groceries purchased (where consumers have access to carbon labels). Such notices could extend to other areas, alerting users to factors such as air and water pollution, resource depletion, and green packaging.

Media and civic engagement. Media-quality applications could use showing notices to alert people to misleading or biased news sources, both on a case-by-case basis and in their overall news consumption. New tools could gradually introduce alternate points of view, encouraging users to break out of ideological echo chambers. Smart citizen phone apps now allow users to

Media-quality applications could alert people to misleading or biased news sources. New tools could gradually introduce alternate points of view, encouraging users to break out of ideological echo chambers.

develop localized crowdsourced maps revealing problem areas for litter, broken streetlights and windows, vandalism, potholes, and so on. Aptly designed visceral notices could track users' interventions and encourage citizens to increase their levels of civic awareness and engagement on local, national, and international levels, prompting them to take action where help is needed.

Skill maintenance. Our willingness to outsource tedious physical engagements with the external world may lead to a significant loss of everyday skills. GPS mapping and automated driving systems are cases in point. When following the visual or voice directions today's systems offer, users don't need to pay attention to landmarks and therefore may not be able to recall routes taken. Visceral notices offer a potential corrective. An AI-enabled system could include a setting that would mimic the way a person on the street might give directions but enhanced by 3D images of key landmarks and points of reference where turns must be made. This would give users the option of orienting themselves to their surroundings and relying on their own memory to reach their destinations instead of mechanically following voice commands as they are given. Other designs could encourage drivers to stay alert and to maintain their driving skills instead of becoming overly reliant on automated driving systems.

Personal edification. Ultimately, what aptly designed visceral notice environments can provide are AI systems that act less like objects and more like friends that help users develop to their fullest potential. Consider the capacity of AI systems to encourage greater discernment in domains such as the arts, cuisine, fashion, and entertainment. Instead of exposing people to whatever products they may react most impulsively to, as recommendation engines often do, they could show alternatives with high-quality ratings based not merely on popularity but also on a blend of expert opinion and personal and shared social preferences. Some services such as Netflix already provide such distinctions, but without a feature showing how the user's overall viewing choices and screen time map to the quality ratings.

AI-assisted platforms provide consumers with extraordinarily powerful tools for controlling and managing their daily lives, activities, and interactions. Such technology, if designed carefully and conscientiously, also holds the power to alter human behavior for the better on a massive scale. But if designed shortsightedly, with few if any features for counteracting its own negative habit-forming effects, it could instead foster passivity, dependency, ignorance, and vulnerability. Applied to millions, these forces undermine the systems of liberal democracy and capitalism.

It is essential that companies working in this area formulate clear and cogent design strategies to allow customers to make informed choices regarding their own patterns of online behavior. The ones that do will establish stronger relationships with their customers while playing a key role in optimizing collective well-being by safeguarding personal agency.

IV

Future Thoughts

17

Building a Robotic Colleague with Personality

Guy Hoffman, interviewed by Frieda Klotz

Anxieties about whether machines will take our jobs will soon be a thing of the past. Robots are already here, adding new dimensions to the way we live and function, and researchers are exploring how to create intelligent machines that work better with us as opposed to taking our place. Guy Hoffman, assistant professor and Mills Family Faculty Fellow in the Sibley School of Mechanical and Aerospace Engineering at Cornell University in Ithaca, New York, is studying how a working robot's behavior can influence its human colleagues. The robots he designs lean forward to show they are listening to human interlocutors, and when they hear music, they nod in response to the beat. Hoffman's work indicates that subtle changes in a robot's actions have a positive effect on the humans around it. *MIT Sloan Management Review* spoke with him about his research to probe what his findings imply for managing human-robot teams.

***MIT Sloan Management Review*: Why do robots need to understand human body language and guess our intentions?**

Guy Hoffman: Robots have traditionally been designed to carry out preprogrammed behaviors. But increasingly, researchers in my field are thinking about modeling human intentions and taking human needs into account. In the past, a robot would perform a fixed action and the human had to adapt to it, but now we want the robot and the human to adapt mutually to each other. For this to happen, the robot has to solve a lot of really hard problems that for us are almost intuitive, which is to guess what we're trying to do or what personality type we have or which mood we might be in. When we encounter people at work, we very quickly make judgments about their personalities and change our behavior accordingly. Having a robot able to do this is crucial if it's to become a similarly good team member.

What part do emotions play in human-robot interactions?

Hoffman: Robots have the capacity to affect our behavior emotionally in that they're using a physical body, they're sharing space with us, they're moving in our surroundings. What I've been looking at is how robots use their bodies to express their intentions, to express what they're trying to do, and therefore affect people emotionally. We've found that when a robot uses human body language, it enables the people interacting with it both to be more effective in what they're doing and to enjoy the interaction and gain psychological benefits from it. I believe that body language and the way that we think with our bodies and through our bodies is the fastest way to our hearts.

So how would these benefits play out in the workplace?

Hoffman: In one study, I looked at robots that solely behaved like robotic tools and the interactions they had compared with robots that were more socially expressive. In the group with the more traditional robots, participants told the machines what to do, and they did it. In the other group, the robots would start moving before they'd been told what to do, and they'd start to help even before they were sure what the person wanted. People who worked with this second group of robots got into a kind of a dance, a back and forth—everybody was moving at the same time and getting things done, even though the robot was taking more chances and would sometimes make mistakes. The results showed that people felt this robot was a better team member and had more commitment to the joint activity. When participants just ordered the robot around, they felt it was lazy; it didn't take the initiative and wasn't a good team member or committed to the team.

In a later study, we had robots listen to people's stories. In that situation, participants weren't working with the robot but were using it to get something off their chest. We specifically chose stories that were negative or traumatic, and the robot would nod at the right moment or lean forward to show that it listened and understood. The participants liked this robot more than one that seemed distracted or didn't react at all. They thought it was smarter. Afterwards they even felt more confident about themselves when going into a stressful task. This showed that people can reap psychological benefits when a robot uses its body even in a very, very small way to show empathy.

And a third project was a musical collaboration in which a pianist and a robot played a piece together. In one case, the robot just played; it didn't exhibit any social expression. In the other, the robot joined the music socially by nodding its head and moving to the beat, looking at the pianist and then back, looking down when it was focused and then up when it was ready for more information. When we asked people to rate the music, they thought it sounded better when the robot used social behaviors than when it acted more mechanically. They thought the musicians were on the same page, as more of a duo than two separate layers. This shows us that body language is not just the icing on the cake but actually changes the taste of the cake. And the same sorts of benefits hold for robots' cooperation and companionship with humans.

Human behavior is so complex. How do you decide how robots should act out?

Hoffman: The way I think about it is very inspired by the arts, from my experience studying theater and playing jazz. Actors have developed tricks for turning what is essentially a very schematic and structured activity into one that appears natural and spontaneous. On stage, good actors look very natural; it looks as though the lights are on in the character's brain. One thing they do is begin a movement before they know where it's going to end. It's called the impulse versus the cue, when they go to speak before their line occurs.

And then there's improvisation, something I looked at in the musical domain, but I feel like it has its place anywhere. I think robots that could improvise at your fast-food restaurant chain would be more fluent and therefore better robotic team members—which will in turn make them more acceptable to the people working with them.

You describe effective human-robot teams as having what you call "collaborative fluency." Is that what you're talking about here?

Hoffman: When I started looking at robots that could anticipate what you wanted to do, I focused on robot-human teams that were building simulated cars together. A surprising finding that emerged was that even though people felt that this sort of robot was much better and smarter at doing a task, it took the team the same amount of time to finish the task. (Though in some of our research, they actually worked more quickly—it depends on the task.)

That's when I came up with the concept of collaborative fluency. What was different about the interaction was that there was a stronger sense of teamwork, a sense that everybody was doing their part and committed to the same ends.

Think about how you interact with Siri or Google Voice or Alexa or the latest intelligent agent: It's very much a back and forth, almost like a chess game, with one move following another. "Can I do this?" I get a response. But if you and I are talking about something we're engaged in, if we're a team that's brainstorming about something, that's not how our conversation goes. You interrupt me, I interrupt you, we build on each other, we complete each other's sentences.

Collaborative fluency occurs when you have this sense of two or more people just rising together like a great football team or a world-class ballet. It's almost like one mind moving together. It's a very subjective feeling, but we're trying to deconstruct it into a mathematical computational model. I believe this is going to be the difference between robots that are going to be a joy to work with and robots that will be annoying to work with and just make you feel as though you have [just] another job.

What advantages would this sort of robot hold for businesses?

Hoffman: I would imagine that companies are interested in their employees' well-being. It would probably also have tangible outcomes for retention and turnover.

If we're building technology that interacts with people, we should think about human values and the well-being of the people working with these robots. In the end, we're building technology to improve our lives. There's no point in just making the world incredibly efficient and depressing.

There's a lot of anxiety about the roles that robots will play in our workplaces. Presumably having more agreeable robots will make the shift an easier one?

Hoffman: Right. Obviously, robots are going to replace people in some cases—it would be naive to think that's not part of the story—but in many cases, and we're already seeing this, robots and humans are working together. I was at a Ford automotive plant recently and saw robots and people producing the same cars, and in my view, we will soon see this in a lot of settings. In addition, robots will be able to use data more effectively and make independent decisions so that a lot of the lower-level decision-making can be done autonomously, and only the higher-level decisions transferred to human workers. We can see this already in collaborative surgery, where the robot may stitch up a wound but doesn't need to be told exactly what stitch to use and how to space it.

We also see more robots coming into retail right now—offering customers promotions, for instance. They're not only going to be facing customers, though; they're also going to be working with human salespeople and human stock-workers. I believe that we'll see this in the restaurant business and in fast-food restaurants, too, with robots working alongside human kitchen workers.

In all these places, we want to create a situation that is beneficial to the people working with the robots by designing robots that support their psychological well-being. I believe that the way these robots interact socially and communicate is going to be a key factor to make this more a utopian and less a dystopian future.

Contributors

Cynthia M. Beath is professor emerita at the University of Texas McCombs School of Business in Austin. Her research focuses on how organizations get value from investments in IT. She is the coauthor (with Jeanne W. Ross and Martin Mocker) of *Designed for Digital: How to Architect Your Business for Sustained Success* (MIT Press, 2019).

Megan Beck is CIO at OpenMatters, a machine learning company, and a research fellow at the SEI Center for Advanced Studies in Management at the Wharton School of Business.

Joe Biron is chief technology officer, internet of things, at PTC, a global software company based in Boston, Massachusetts.

Erik Brynjolfsson is the Schussel Family Professor of Management Science at the MIT Sloan School of Management and director of the MIT Initiative on the Digital Economy (IDE).

Jacques Bughin is a senior partner in the Brussels office of McKinsey & Co. and a director of the McKinsey Global Institute.

Rumman Chowdhury is a data scientist and social scientist and Accenture's global lead for responsible AI.

Paul R. Daugherty is Accenture's chief technology and innovation officer.

Thomas H. Davenport is the President's Distinguished Professor of Information Technology and Management at Babson College in Wellesley, Massachusetts, as well as a fellow at the MIT IDE and a senior adviser to Deloitte's Analytics and Cognitive practice. He is the author of *The AI Advantage: How to Put the Artificial Intelligence Revolution to Work* (MIT Press, 2018).

Chris DeBrusk is a partner in the financial services and digital practices of Oliver Wyman, a global management consulting firm.

Berkeley J. Dietvorst is assistant professor of marketing at the University of Chicago Booth School of Business.

Janet Foutty is chair and CEO of Deloitte Consulting LLP.

James R. Freeland is the Sponsors Professor of Business Administration in the technology and operations area at Darden School of Business.

R. Edward Freeman is a professor of strategy, ethics, and entrepreneurship at the Darden School of Business at the University of Virginia.

Julian Friedland is an assistant professor of ethics at Trinity Business School at Trinity College Dublin.

Lynda Gratton is a professor of management practice at London Business School and director of the school's Human Resource Strategy in Transforming Companies program. She is coauthor of *The 100-Year Life: Living and Working in an Age of Longevity* (Bloomsbury, 2016).

Francis Hintermann leads Accenture Research.

Guy Hoffman is assistant professor and Mills Family Faculty Fellow in the Sibley School of Mechanical and Aerospace Engineering at Cornell University.

Vivek Katyal is the global data risk leader and leads the advisory analytics service area in the United States for Deloitte.

David Kiron is executive editor of *MIT Sloan Management Review.*

Frieda Klotz is a journalist who writes about technology and health care.

Jonathan Lang is lead principal business analyst at PTC.

Barry Libert is CEO of OpenMatters and a senior fellow at Wharton's SEI Center.

Paul Michelman is editor in chief of *MIT Sloan Management Review.*

Sam Ransbotham is an associate professor of information systems at the Carroll School of Management at Boston College and the *MIT Sloan Management Review* guest editor for the Artificial Intelligence and Business Strategy Big Idea Initiative.

Daniel Rock is a PhD candidate at MIT Sloan School of Management and researcher at IDE.

Jeanne W. Ross is a principal research scientist at MIT Sloan's Center for Information Systems Research. She is the coauthor (with Cynthia M. Beath and Martin Mocker) of *Designed for Digital: How to Architect Your Business for Sustained Success* (MIT Press, 2019) and also writes a regular column for *MIT Sloan Management Review* about digital business management issues.

Eva Sage-Gavin leads the Talent and Organization practice for Accenture.

Chad Syverson is the Eli B. and Harriet B. Williams Professor of Economics at the University of Chicago Booth School of Business.

Monideepa Tarafdar is a professor of information systems and codirector of the Centre for Technological Futures at Lancaster University Management School in the United Kingdom. Her research focuses on how digital technologies transform organizations and societies.

Gregory Unruh is the Arison Group Endowed Professor of Doing Good Values at George Mason University in Fairfax, Virginia, and the *MIT Sloan Management Review*'s guest editor for the Sustainability Big Ideas Initiative.

Madhu Vazirani is a director with Accenture Research.

H. James Wilson is managing director of IT and business research at Accenture Research.

Notes

Introduction

1. S. Ransbotham, "Accelerate Access to Data and Analytics with AI," Big Idea: Artificial Intelligence and Business Strategy, *MIT Sloan Management Review*, August 22, 2017, https://sloanreview.mit.edu/article/accelerate-access-to-data-and-analytics-with-ai/.
2. S. Ransbotham, "Improving Customer Service and Security with Data Analytics," Big Idea: Competing with Data & Analytics, *MIT Sloan Management Review*, July 25, 2017, https://sloanreview.mit.edu/article/improving-customer-service-and-security-with-data-analytics/.
3. S. Ransbotham, "Get the Most Out of AI Today," Artificial Intelligence and Business Strategy, *MIT Sloan Management Review*, July 25, 2018, https://sloanreview.mit.edu/article/get-the-most-out-of-ai-today/.

Chapter 3: Using Artificial Intelligence to Promote Diversity

1. L. Hardesty, "Study Finds Gender and Skin-Type Bias in Commercial Artificial Intelligence Systems," MIT News Office, February 11, 2018, http://news.mit.edu/2018/study-finds-gender-skin-type-bias-artificial-intelligence-systems-0212.
2. E. T. Israni, "When an Algorithm Helps Send You to Prison," *New York Times*, October 26, 2017, https://www.nytimes.com/2017/10/26/opinion/algorithm-compas-sentencing-bias.html.
3. L. Camera, "Women Can Code—as Long as No One Knows They're Women," *US News & World Report*, February 18, 2016, https://www

.usnews.com/news/blogs/data-mine/2016/02/18/study-shows-women-are-better-coders-but-only-when-gender-is-hidden.
4. M. Muro, A. Berube, and J. Whiton, "Black and Hispanic Underrepresentation in Tech: It's Time to Change the Equation," Brookings Institution, March 28, 2018, https://www.brookings.edu/research/black-and-hispanic-underrepresentation-in-tech-its-time-to-change-the-equation/.
5. "About Us," 2019, https://girlswhocode.com/about-us/.
6. F. Dobbin and A. Kalev, "Why Diversity Programs Fail," *Harvard Business Review* 94, no. 7/8 (July–August 2016), https://hbr.org/2016/07/why-diversity-programs-fail.
7. R. Locascio, "Thousands of Sexist AI Bots Could Be Coming. Here's How We Can Stop Them," *Fortune*, May 10, 2018, http://fortune.com/2018/05/10/ai-artificial-intelligence-sexism-amazon-alexa-google/.
8. "Inclusive Design," https://www.microsoft.com/design/inclusive/.
9. T. Halloran, "How Atlassian Went From 10% Female Technical Graduates to 57% in Two Years," *Textio Word Nerd* (blog), December 12, 2017, https://textio.ai/atlassian-textio-81792ad3bfbf?gi=166f8343c9ee.
10. C. DeBrusk, "The Risk of Machine-Learning Bias (and How to Prevent It)," *MIT Sloan Management Review*, March 26, 2018, https://sloanreview.mit.edu/article/the-risk-of-machine-learning-bias-and-how-to-prevent-it/.
11. J. Zou and L. Schiebinger, "AI Can Be Sexist and Racist—It's Time to Make It Fair," *Nature*, July 12, 2018, https://doi.org/10.1038/d41586-018-05707-8.
12. D. Bass and E. Huet, "Researchers Combat Gender and Racial Bias in Artificial Intelligence," *Bloomberg Technology*, December 4, 2017, https://www.bloomberg.com/news/articles/2017-12-04/researchers-combat-gender-and-racial-bias-in-artificial-intelligence.
13. B. Lovejoy, "Sexism Rules in Voice Assistant Genders, Show Studies, but Siri Stands Out," February 22, 2017, *9to5Mac*, https://9to5mac.com/2017/02/22/siri-sexism-intelligent-assistants-male-female/.
14. J. Elliot, "Let's Stop Talking to Sexist Bots: The Future of Voice for Brands," *Fast Company*, March 7, 2018, https://www.fastcompany.com/40540892/lets-stop-talking-to-sexist-bots-the-future-of-voice-for-brands.
15. S. Paul, "Voice Is the Next Big Platform, Unless You Have an Accent," *Wired*, March 20, 2017, https://www.wired.com/2017/03/voice-is-the-next-big-platform-unless-you-have-an-accent/.

Chapter 11: Using AI to Enhance Business Operations

1. ECC applications are distinct from other kinds of enterprise software in that AI tools, rather than human deduction, are used to figure out what logic will optimize business outcomes. AI software tools apply computational and analytical techniques, such as neural network analysis, machine learning, and Bayesian statistics, to large sets of structured and unstructured data to create AI algorithms that will classify, cluster, predict, and match patterns. These algorithms become part of the logic of the ECC application.

2. J. Bughin and E. Hazan, "Five Management Strategies for Getting the Most from AI," *MIT Sloan Management Review*, September 19, 2017, https://sloanreview.mit.edu/article/five-management-strategies-for-getting-the-most-from-ai/.

3. M. Tarafdar, C. M. Beath, and J. W. Ross, "Enterprise Cognitive Computing Applications: Opportunities and Challenges," *IEEE IT Professional* 19, no. 4 (August 2017): 21–27.

4. S. Ransbotham, D. Kiron, P. Gerbert, and M. Reeves, "Reshaping Business with Artificial Intelligence," *MIT Sloan Management Review* research report, September 6, 2017, https://sloanreview.mit.edu/projects/reshaping-business-with-artificial-intelligence/.

5. S. Norton, "Machine Learning at Scale Remains Elusive for Many Firms," *Wall Street Journal*, April 27, 2018, https://blogs.wsj.com/cio/2018/04/27/machine-learning-at-scale-remains-elusive-for-many-firms/; Bughin and Hazan, "Five Management Strategies"; and Ransbotham et al., "Reshaping Business."

Chapter 16: AI Can Help Us Live More Deliberately

1. R. Wollheim, *The Thread of Life* (New Haven, CT: Yale University Press, 1984).

2. M. J. Sandel, "The Case against Perfection: What's Wrong with Designer Children, Bionic Athletes, and Genetic Engineering," *Atlantic Monthly*, April 2004, 1–11; and S. Vallor, *Technology and the Virtues: A Philosophical Guide to a Future Worth Wanting* (Oxford: Oxford University Press, 2016).

3. B. Frischmann and E. Selinger, *Re-Engineering Humanity* (Cambridge: Cambridge University Press, 2018).

4. G. Lukianoff and J. Haidt, *The Coddling of the American Mind: How Good Intentions and Bad Ideas Are Setting Up a Generation for Failure* (New York: Penguin Random House, 2018).

5. D. Kahneman, *Thinking, Fast and Slow* (New York: Farrar, Straus and Giroux, 2011).

6. J. Friedland and B. M. Cole, "From Homo-economicus to Homo-virtus: A System-Theoretic Model for Raising Moral Self-Awareness," *Journal of Business Ethics* 155, no. 1 (March 2019): 191–205.

7. Brett Frischmann and Evan Selinger describe the six forms of disengagement in *Re-Engineering Humanity*.

8. N. Carr, *The Glass Cage: How Our Computers Are Changing Us* (New York: W.W. Norton & Co., 2015).

9. W. Knight, "Socially Sensitive AI Software Coaches Call-Center Workers," *MIT Technology Review*, January 31, 2017, https://www.technologyreview.com/s/603529/socially-sensitive-ai-software-coaches-call-center-workers/.

10. P. Lin, "Why Ethics Matters for Autonomous Cars," in M. Maurer, J. C. Gerdes, B. Lenz, and H. Winner, eds., *Autonomous Driving: Technical, Legal, and Social Aspects* (Berlin: Springer Open, 2016), 69–85.

11. Sandel, "The Case against Perfection."

12. Carr, *The Glass Cage*.

13. R. H. Thaler and C. R. Sunstein, *Nudge: Improving Decisions about Health, Wealth, and Happiness* (New Haven, CT: Yale University Press, 2008).

14. K. G. Volpp, A. B. Troxel, S. J. Mehta, L. Norton, J. Zhu, W. Wang, et al., "Effect of Electronic Reminders, Financial Incentives, and Social Support on Outcomes after Myocardial Infarction: The HeartStrong Randomized Clinical Trial," *JAMA Internal Medicine* 177, no. 8 (August 2017): 1093–1101.

15. Vallor, *Technology and the Virtues*, 161.

16. B. Lamm, H. Keller, J. Teiser, H. Gudi, R. D. Yovsi, C. Freitag, et al., "Waiting for the Second Treat: Developing Culture-Specific Modes of Self-Regulation," *Child Development* 89, no. 3 (June 2018): e261–e277.

17. T. W. Watts, G. J. Duncan, and H. Quan, "Revisiting the Marshmallow Test: A Conceptual Replication Investigating Links Between Early

Delay of Gratification and Later Outcomes," *Psychological Science* 29, no. 7 (May 2018): 1159–1177.

18. A. Duckworth and J. J. Gross, "Self-Control and Grit: Related but Separable Determinants of Success," *Current Directions in Psychological Science* 23, no. 5 (October 2014): 319–325.

19. Vallor, *Technology and the Virtues*, 163.

20. Friedland and Cole, "From Homo-Economicus to Homo-Virtus."

21. K. Aquino and A. Reed II, "The Self-Importance of Moral Identity," *Journal of Personality and Social Psychology* 83, no. 6 (December 2002): 1423–1440.

22. S. Bowles, *The Moral Economy: Why Good Incentives Are No Substitute for Good Citizens* (New Haven, CT: Yale University Press, 2016).

23. K. Hwang and H. Kim, "Are Ethical Consumers Happy? Effects of Ethical Consumers' Motivations Based on Empathy versus Self-Orientation on Their Happiness," *Journal of Business Ethics* 151, no. 2 (2018): 579–598.

24. J. Sadowski, S. G. Spierre, E. Selinger, T. P. Seager, E. A. Adams, and A. Berardy, "Intergroup Cooperation in Common Pool Resource Dilemmas," *Science and Engineering Ethics* 21, no. 5 (October 2015): 1197–1215.

25. R. B. Cialdini, C. A. Kallgren, and R. R. Reno, "A Focus Theory of Normative Conduct: A Theoretical Refinement and Re-Evaluation of the Role of Norms in Human Behavior," *Advances in Experimental Social Psychology* 24 (December 1991): 201–234.

26. J. Sadowski, T. P. Seager, E. Selinger, S. G. Spierre, and K. P. Whyte, "An Experiential, Game-Theoretic Pedagogy for Sustainability Ethics," *Science and Engineering Ethics* 19, no. 3 (September 2013): 1323–1339.

27. A. Gopaldas, "Marketplace Sentiments," *Journal of Consumer Research* 41, no. 4 (December 1, 2014): 995–1014.

28. Hwang and Kim, "Are Ethical Consumers Happy?"

29. L. Columbus, "2017 Roundup of Internet of Things Forecasts," *Forbes*, December 10, 2017, https://www.forbes.com/sites/louiscolumbus/2017/12/10/2017-roundup-of-internet-of-things-forecasts/.

Index